Human Factors and Safety Culture

This title explores human behaviour in the context of workplace safety and risk management. Focused on understanding how people detect, interpret and respond to danger and how leaders can put safety at the heart of their organizations' culture, it draws on the latest insights from disciplines such as cognitive science, neuroscience, psychology, and sociology. Integrating traditional and emerging perspectives in the field of Occupational Health and Safety, this book delivers both a vision and the tools to elevate safety as a core organizational value able to motivate and anchor safe behaviours and reinforce safety-oriented leadership.

Written to include practical frameworks and clear examples, it addresses the cognitive processes, including perception, attention, and memory, that influence individuals' judgement and decision-making at work as well as spontaneous behaviour. Readers will discover how biases, emotions, and underlying values play a role in shaping attitudes towards safety, providing a fresh perspective on emotional intelligence and behavioural motivation. Through a "Toolbox-style" section, filled with actionable techniques that can be applied to any workplace, readers gain strategies to implement these insights immediately, helping to embed safety as a shared cultural value.

Additional sections, such as "Did you know?" and "Focus on...", present surprising findings and deeper dives into key topics, revealing real-world applications. The reader will develop a good understanding of the key theories and practices behind safety culture at work that can be made applicable to any industry. *Human Factors and Safety Culture: How Leaders Can Influence Behaviours for Good* is designed for those in occupational health and safety, including current and aspiring safety leaders, HR and operations managers, and anyone involved in shaping a positive organizational workplace culture.

Human Factors and Safety Culture

How Leaders Can Influence Behaviours for Good

Eduardo Blanco-Muñoz

CRC Press
Taylor & Francis Group
Boca Raton London New York

CRC Press is an imprint of the
Taylor & Francis Group, an **informa** business

Designed cover image: Shutterstock

First edition published 2026
by CRC Press
2385 NW Executive Center Drive, Suite 320, Boca Raton FL 33431

and by CRC Press
4 Park Square, Milton Park, Abingdon, Oxon, OX14 4RN

Original French version "Facteur humain et Culture sécurité : Science et pratique du leadership en sécurité" published 2024 by CNPP Editions, Saint-Marcel, France.

CRC Press is an imprint of Taylor & Francis Group, LLC

ISBN: 978-1-041-02012-7 (hbk)
ISBN: 978-1-041-01936-7 (pbk)
ISBN: 978-1-003-61734-1 (ebk)

DOI: 10.1201/9781003617341

Typeset in Times
by codeMantra

Contents

Figures

Foreword

Safety is not a box to tick or a checklist to follow – it's a reflection of the values we hold and the culture we create in our workplaces. It's about more than compliance or meeting standards; safety is about shaping the way people think, feel, and act every single day on the job. It's about embedding safety into the very DNA of your organization.

In *Human Factors and Safety Culture: How Leaders Can Influence Behaviours for Good*, Eduardo Blanco-Muñoz addresses the essence of what makes safety work – or fail. He challenges traditional views of safety leadership, urging us to move beyond rules and protocols and focus on the human elements that drive behaviour. This book is not just about theory; it's about how leaders can directly influence the culture of their organizations and foster behaviours that ensure safety is at the heart of everything.

RETHINKING SAFETY LEADERSHIP

Eduardo Blanco-Muñoz's approach is built on the understanding that safety leadership isn't about titles or formal authority – it's about behaviours and influence. This resonates deeply with my own journey across nearly three decades of working with leaders around the world. From the factory floors of Southeast Asia to corporate boardrooms in North America, I've seen first-hand the profound impact that leadership has on safety culture.

Good safety leadership isn't about issuing commands or handing out procedures; it's about inspiring commitment, trust, and responsibility. A strong safety culture thrives when leaders model the behaviours they want to see in their teams – when they connect authentically with people and communicate the importance of safety in ways that resonate on a personal level. Eduardo Blanco-Muñoz captures these dynamics beautifully, offering a roadmap for leaders who want to drive meaningful change.

At its core, this book is structured around three key themes: values, beliefs, and practices. These are the levers leaders must understand and master to create lasting impact.

The first theme, values, is the foundation of any safety culture. As Eduardo Blanco-Muñoz rightly emphasizes, leaders must demonstrate through their actions that safety is non-negotiable. It's not enough to say safety is a priority – actions speak louder than words. Eduardo Blanco-Muñoz provides a clear framework for aligning values with behaviours, ensuring that safety becomes intrinsic to the organization's culture rather than a fleeting priority.

The second theme, beliefs, delves into the cognitive and emotional factors that shape our decisions. Understanding why people behave the way they do in safety-critical situations is essential for any leader looking to influence those behaviours. Eduardo Blanco-Muñoz explores biases, heuristics, and emotional drivers with clarity, showing how these factors can either support or undermine safety efforts. For me, this section reflects an often-overlooked truth: what people believe about risk and safety shapes how they act in the moment – and how they approach work overall.

Finally, this book transitions into practices – practical strategies that leaders can implement to make safety a lived experience, not just an aspiration. The "Toolbox" sections offer tools and techniques that are immediately actionable, grounded in research and Eduardo Blanco-Muñoz's extensive experience. These strategies are the kind of practical, results-driven methods I've seen work across organizations around the world in industries such as construction, mining, manufacturing, health-care, and more.

LESSONS FROM A GLOBAL PERSPECTIVE

In my own career, I've worked with thousands of leaders in over 130 countries, spanning industries and cultures as diverse as their approaches to safety. What I've learned is that while safety challenges vary, the principles of effective safety leader-ship remain universal.

Eduardo Blanco-Muñoz's book resonates because it speaks to those universal principles while recognizing the complexity of human behaviour. It bridges the gap between understanding individual factors – like perception and cognition – and the broader organizational systems that shape culture. This integration of human factors and safety culture is precisely what leaders need to drive progress in today's rapidly changing world.

As someone who has spent years teaching safety leadership and culture at some of the world's leading business schools, I see first-hand how transformative these concepts can be. Eduardo Blanco-Muñoz's work brings them to life in a way that is accessible yet deeply insightful. Whether you are an experienced safety professional, a student, or a manager navigating safety for the first time, this book offers a fresh perspective and practical guidance.

WHY THIS BOOK MATTERS

Human Factors and Safety Culture is more than just a theoretical exploration – it's a practical guide that can transform how safety is approached within organiza-tions. Eduardo Blanco-Muñoz presents a clear and actionable framework that directly links human factors to safety culture, illustrating how leadership actions can shape behaviour and foster meaningful change. This book challenges common assumptions about human behaviour, delving into the biases and emotional driv-ers that influence decision-making and offering a deeper understanding of why people don't always follow safety protocols. Its practical value is undeniable, with a "Toolbox" filled with real-world strategies that leaders can implement immediately to drive results. Grounded in rigorous research, with over 200 academic refer-ences, this book balances scientific depth with accessibility. By pushing us to move beyond traditional safety practices, Eduardo Blanco-Muñoz provides a fresh and forward-thinking perspective on what it takes to embed safety into organizational culture effectively.

A CALL TO RETHINK, REINVENT, AND LEAD

If you're ready to move beyond surface-level safety initiatives and create a culture where safety truly takes root, this book offers the clarity and tools to make it happen. For those frustrated by inertia or the same recurring barriers to progress, it provides fresh perspectives and actionable solutions to break through and lead real change.

In *Human Factors and Safety Culture*, Eduardo Blanco-Muñoz has created a resource that is both insightful and practical. It's a wake-up call for leaders, organizations, and safety professionals alike, and I'm delighted to recommend it to you.

Professor Dr Andrew Sharman
Chief Executive
International Institute of Leadership & Safety Culture
www.iilsc.com

Preface

This book is addressed to all safety leaders and to those who want to become one. Is it meant for you?

Certainly. Here's why.

To be a safety leader is neither a title nor a function: it is an attitude that translates into specific, observable behaviours that inspire others to act safely. We will see along these pages that safety leadership and the related behaviours are grounded on values, beliefs, and practices that we can characterize and promote – and you will learn to do just that.

When it comes to values, your humble servant is sure that having chosen this book, yours are already compatible with the role: preventing injuries, illness, and suffering is important enough to you to invest time and money on the topic. You will find here the tools to successfully share and grow these values within your organization.

About beliefs, we will investigate a series of phenomena that make people at work, individually and collectively, behave as they do. Many of these factors are poorly understood, and maybe you will review some of your own beliefs about your behaviours and those of your co-workers. Myths will be exposed, and hidden nuggets of wisdom will be unveiled. The "Did you know?" sections will present the counter-intuitive information that may surprise you. The "Focus on..." sections will further develop some relatively sophisticated concepts. You will even find some "Tests" so you can experience first-hand several surprising tricks that you will be able to share.

Finally, here come the practices. You can consider this book as a journey to explore and (re)discover the science of human behaviour at work, but above all, it intends to be practical. A series of techniques and methods that you can use right away is presented in the form of a "Toolbox" distilled across this book. These tools will help you to implement in your organization the path to anchoring safety as a fundamental element of its culture.

The academics well versed in one or more of the topics presented here will excuse the theoretical simplifications required by this pragmatic (and hopefully, even entertaining) approach: further explanations are accessible via the bibliography provided. However, hasty readers can navigate the chapters by focusing on the sections "Key Learning Points to the Safety Leader" and then dwell a bit longer on the subjects that may capture their interest. In any case, please note that the successive chapters form a whole that unfolds progressively by stepping on the concepts exposed in the previous sections.

To the former and to the latter, thank you. Whatever your experience and character, your organizations need your enlightened contributions. Whatever your hierarchical level and your function, they need you to understand and be able to guide the human factors at work so as to put safety front and centre in your culture. I hope you will find in these pages some useful keys.

About the Author

Eduardo Blanco-Muñoz is an occupational health and safety practitioner, professor, and author based in Paris, France. He has been working in HSE functions for a quarter of a century. His career started in the chemicals industry as a QHSE Officer, and then as a Manager at several Seveso "upper threshold" sites, which evolved to a role in senior management positions in a series of large multinational groups in sectors as diverse and challenging as optics and medical devices, energy, aerospace, logistics and transportation, and construction and civil engineering. His scientific and technical training began in Spain, continued in the United Kingdom, and was completed in France: he holds three master's degrees in Environmental Sciences, Environmental Engineering, and Risk Management. He is a regular key speaker at conferences and corporate events, and after teaching behavioural safety and safety culture at Sorbonne Paris Nord University, he is an affiliated professor at the French *Conservatoire National des Arts et Métiers*.

1 Introduction
Human Performance at the Heart of Prevention

"The fool doth think he is wise, but the wise man knows himself to be a fool".

William Shakespeare

HUMANS AND SAFETY

"No tendency is quite so strong in human nature as the desire to lay down rules of conduct for other people".

William Howard Taft

What is the role of the worker when it comes to occupational safety? Besides being a potential victim, many safety practitioners still consider people and the "human factors" that they embody as one of the three possible causes of an accident. Indeed, when a cause analysis is performed, we often hear about (1) technological factors, (2) organizational factors, and (3) human factors.

We keep on hearing that most accidents are caused by human shortcomings, errors, deviations, and violations, a paradigm that became extremely popular over the last century following the findings of Heinrich. He stated that the vast majority of safety issues derived from "man-failure". Even though Heinrich's writings have been grossly oversimplified – and his research extensively criticized as unscientific and outdated – this idea still impregnates the mindset of the profession.

DID YOU KNOW? HUMANS AS THE WEAKEST LINK: A 100-YEAR-OLD OPINION

Herbert W. Heinrich (1886–1962) published *Industrial Accident Prevention, A Scientific Approach* in 1931 on the basis of the empirical findings that he made while working in the Engineering and Inspection Division of an insurance company, a job that gave him access to thousands of accident reports (Heinrich, 1931).

Other than what became known as *Heinrich's law* – a questionable ratio which says that for every accident that causes a severe injury in a workplace, there are 29 accidents that cause just minor injuries and 300 that cause no injury – he figured that 88% of accidents were caused by "man-failure".

DOI: 10.1201/9781003617341-1

Although he also insisted on the need for hazard control and a hundred pages of his book were dedicated to machine guarding, the main message that many kept in mind is that human fallibility and "unsafe acts" (which he never really addressed as such) are the main cause of accidents in the workplace.

It is to be noted that Heinrich's approach, methods, and results have been widely refuted since the 1980s (Manuele, 2003).

This book intends to further elaborate on the new vision that parts with Heinrich's postulates. The very concept of "human error", as it has traditionally been understood, has been challenged, and there is a growing consensus on the limits of the practical applications of such an ambiguous notion (Hollnagel and Amalberti, 2001). At the very least, we can agree that considering human factors in this old-fashioned light is not just inaccurate; it is counterproductive for at least three reasons:

For starters, this assumption implies that a single factor can cause an accident, which is not true. Even with the simplest causal models, we can only imagine an accident as the result of a combination of different factors at a given moment. It is extremely rare to find examples of situations where an individual's error could trigger significant losses on its own. As a matter of fact, a system where such an eventuality was possible would be poorly designed, if only because it failed to consider the fallibility of a person in charge of a critical task. Thus, we should be talking about the combination of at least two factors: an operator's error and a design deficiency. In the 1920s, when the accidents analysed by Heinrich took place (more than a century ago!), the technological and organizational means to ensure worker safety were in their infancy in most organizations, as was the understanding of human factors. It was only natural that supervisors reporting accidents attributed them to "man-failure".

In addition, such premises implicitly consider the three aforementioned types of factors as causes of trouble. They can be so in certain circumstances, but it is, in general, more appropriate to refer to them as *protection and prevention means*. Instead of the combination of different factors seen as "aggressors", we could then speak of the alignment of a series of *weaknesses* in the protective layers that should allow for a safe operation in a hazardous environment. This is the philosophy behind James Reason's famous "Swiss cheese model" (Reason, 1990).

Finally, it seems naïf to dissociate the human factors from the technological and organizational ones. Such a shortcut certainly gives us a highly practical taxonomy that we will use later in this introductory chapter before developing a more sophisticated view. But it does not stand to reason that technology and organization are independent from human beings. That would amount to saying that the prevention and protection technological means (guarding, sensors, PLCs...) and organizational means (org-charts, SOPs, checklists...) are not imagined, developed, produced, implemented, validated, and maintained by human people.

Taking this into account, the defects that we may find in those elements require a "man-failure" to occur in one of the tasks listed above. Wait... would that not be an "organizational failure" that led to the overlooking of human imperfection? We will try to close this loop in these pages.

FOCUS ON... NATURAL SCIENCES VS SOCIAL SCIENCES

Human factors have been approached from several angles. On one side, the **physicalist trend** historically pushed researchers to investigate the relationship between the worker and the task. The workers' behaviours and techniques are empirically observed with no interest in their individual subjectivity. Imbued with positivism and determinism, this trend favours experimentation and natural sciences to study how people work, aiming to deduce generally applicable "laws".

On the other side, the **psychosociological trend** tries to understand the job by analysing the relationship between the worker and the other players. Assuming the individuals' agency, it is their subjectivity and intersubjectivity that generate their motivation and ultimately trigger their behaviours. The task and the working environment are just considered as, respectively, the job's object and material context. Such subjectivity cannot be measured directly, so this trend favours social sciences as the way to study human factors.

The contributions coming from cognitive sciences, ergonomics, neurosciences, anthropology, and even ethnography have come to add to this picture and offer a more nuanced view on the topic. This book intends to adopt a transversal and holistic approach. First of all, we will take stock of the abundant knowledge about our physiology and the latest discoveries about our brain, which in the last 20 years have shaken up our understanding of our mind thanks notably to the new cerebral imaging techniques. After that, we will move on to explore the psychological dimension – particularly the insights provided by experimental psychology and the research on behavioural economics – and then the social and cultural ones.

Each one of these disciplines sheds additional light on what has become to be known as "safety science" and provides useful tools to the safety practitioner. We will see that their different perspectives are essentially compatible and coherent, and nicely complement each other.

This multidimensional approach is probably the best way to reconcile what Christophe Dejours called the "human being's simultaneous belonging to determinism and freedom", which remains the biggest difficulty in understanding people and their behaviours. As he put it, to inhabit simultaneously the natural world and our own minds places us in a difficult existential situation. We are continuously torn between what we want and what we can, what we desire and what we must. This contradiction affects us in our daily lives and, of course, at work. Human beings cannot be considered unified and coherent, as we are always split and distressed by this internal conflict that we are unable to resolve, and this necessarily requires us to fully consider the psycho-affective dimension when studying human factors (Dejours, 1995–2022).

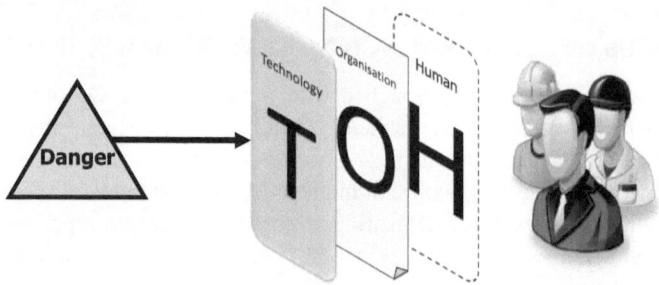

FIGURE 1   The HOT barriers (drawing by author).

HUMAN FACTORS IN A POSITIVE LIGHT

This introduction is relevant because it allows us to envision the human factors at the very heart of occupational safety and, even more importantly, to approach them in a positive way as safety factors and not as causes of accidents, contrary to conventional wisdom.

Let us get back for a minute to the classical dichotomy between technological, organizational, and human factors to illustrate how key and positive this last (but definitely not least) factor really is – we will see later how transversal it is and hence the debatable nature of this dichotomy.

We start by visualizing these three elements not as potential issues but as protective layers that we interpose between danger and the subject that may be damaged or injured.

You will observe in Figure 1 that the first barrier, **technology**, is represented as a solid layer. These are physical protective elements – whether hardware or software – and are usually considered the most reliable. Codes and regulations in many countries have established a hierarchy of control means, and whenever a danger cannot be suppressed, technological means are in principle to be privileged to contain it.

FOCUS ON... THE HIERARCHY OF CONTROL MEANS

The concept of a formal hierarchy of control means was introduced in the 1940s. The widely known inverted triangle, proposed by the National Safety Council in the United States, inspired safety agencies and authorities around the world, who translated it into policies and regulations.

An example is Article 6.2 of European Directive 89/391/EEC, which set in 1989 the following nine "general principles of prevention":

1. Avoiding risks
2. Evaluating the risks which cannot be avoided
3. Combating the risks at source

4. Adapting the work to the individual, especially as regards the design of workplaces, the choice of work equipment, and the choice of working and production methods, with a view, in particular, to alleviating monotonous work and work at a predetermined work rate and to reducing their effect on health
5. Adapting to technical progress
6. Replacing the dangerous with the non-dangerous or the less dangerous
7. Developing a coherent overall prevention policy that covers technology, organization of work, working conditions, social relationships, and the influence of factors related to the working environment
8. Giving collective protective measures priority over individual protective measures
9. Giving appropriate instructions to the workers.

We can see here that, when a risk cannot be avoided or suppressed at its source (principles 1–3), priority is given to technological then organizational measures, and instructions given to workers are mentioned last.

Note that when we refer to *technology* in this book, we mean every device that is used by a worker, from the simplest rod to the most sophisticated computers. We distinguish between *technological* (referring to technology as defined above) and *technical* (a broader concept that will be discussed later).

Prioritizing technological control means seems the natural thing to do when we consider the mechanistic worldview that impregnates our thinking, especially at work and in the safety profession, which is mostly populated by technicians and engineers. As the industrial revolution brought the industrial-scale level of risk that we are facing, we spontaneously look for industrial solutions to mitigate them.

This is certainly legitimate if we consider that these technological solutions are a priori the most dependable ones, as they should be if they are properly engineered, implemented, maintained, etc. But it also reflects the logic that we deploy: hazards in a technological working environment can, and should, be mitigated with technological means. This is not a bad principle, as long as it does not lure us into thinking that technology is all we need and that it will solve everything.

The second line of defence is the **organization** set in and around the job. It is shown in the diagram in Figure 1 as a sheet of paper, representing the usual vector for procedures, SOPs, etc., an image that also tries to convey its brittle nature compared to the technological means presented above.

To consider the organizational dimension of workplace safety seems pretty obvious today, but it was not until quite recently that employers and regulators (not necessarily in that order) took that turn. It was even later that organizations went beyond an approach based on mere *compliance* to one geared to *management systems*. Needless to say, not all organizations have reached that stage.

**DID YOU KNOW? THE QUITE RECENT
STANDARDIZATION OF SAFETY MANAGEMENT**

Whereas trans-organizational standardization of quality practices and methods can be traced back to the 1950s (MIL-Q-9858 standard from the US Department of Defence), the first relevant occupational health and safety management standard was not published until 1996, some 40 years later (BS 8800 from the British Standardization Institute).

Likewise, the first internationally recognized series of standards dedicated to quality management (ISO 9000) appeared in 1987, but we had to wait more than 30 years for the first international standard on occupational health and safety management to arrive, in 2018 (ISO 45001).

Of course, organizations facing major risks like military, law enforcement and emergency response, oil and gas, nuclear power, aerospace, chemicals, and healthcare, to name a few, did not wait for management systems to be standardized to develop and implement their own (Hudson, 2001). It was in such high-stakes environments that the pioneers of safety management were first confronted with the limits of this second layer. Hence, they were the first to formally recognize the key role of front-line operators and to try to enhance their performance.

It was hence in those domains that the **human factors** were originally studied in depth, the "third layer of defence" that we represented as a dotted-lined shape because it is totally intangible: it is invisible to the untrained eye, but it is observable both at the individual level in the workers' techniques and behaviours and at the collective level in the culture of the organization.

The diagram presented in Figure 1 above helps us to understand how, contrary to the negative vision of human beings as a cause of accidents, human factors complete the system of protective layers. But this simplified representation could lead us to believe that this third barrier may be, if not redundant, relatively secondary with regards to technology and organization, or that these three barriers could be considered independent from one another.

HUMAN FACTORS AS TRANSVERSAL FACTORS

In real life, and in the workplace in particular, we do not face a single type of danger (Figure 2). Conceiving and implementing technological and organizational barriers capable of perfectly mitigating all risks is much more complicated than it looks on paper.

The hope of bulletproofing an operation through technology and organization is not realistic. It can even be dangerous, as it generates the illusion of safety, especially among those far away from the frontline. A flashback to Reason and his classic Swiss cheese model reminds us that no protective layer is totally and permanently impervious.

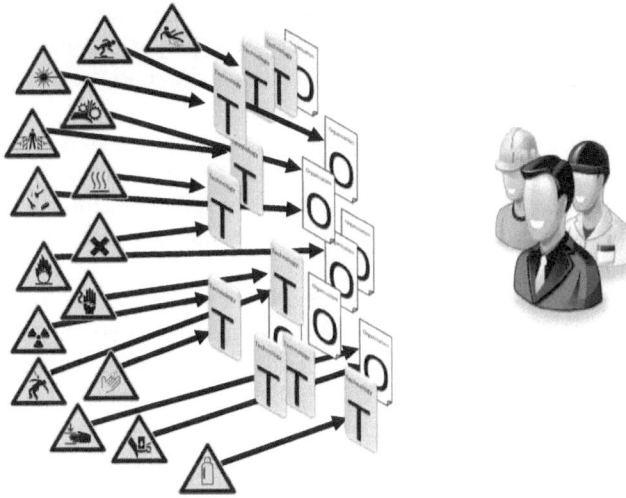

FIGURE 2   Too many dangers (drawing by author).

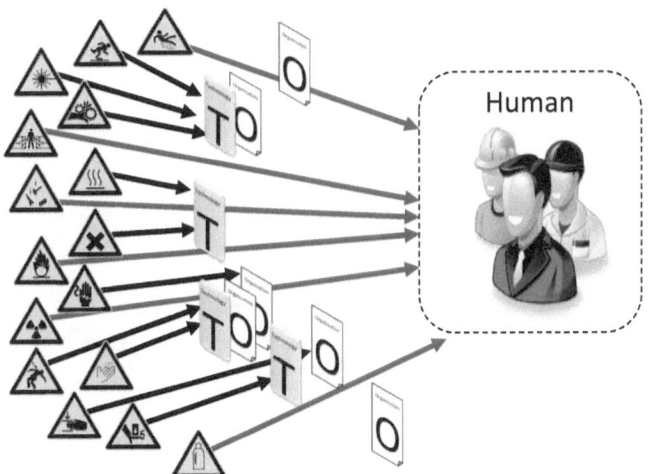

FIGURE 3   Human abilities as the last line of defence (drawing by author).

The first reason why human factors are paramount in occupational safety could be just that this particular protective layer is the only one that we have always on us (Figure 3). Whatever the job and the working environment, the workers will always require a certain degree of flexibility to adapt to the context and perform their duties. Their ability to pick up and quickly interpret weak or ambiguous signs of danger, their capacity to analyse a situation and assess the risks in a dynamic way, and their aptitude to integrate different flows of information and prioritize turn people into assets that have not yet been matched by technology or decision algorithms. Although fallible, our human capabilities are more polyvalent than technological and organizational barriers, which tend to be designed to mitigate a specific type of danger under very precise conditions.

But our unique human abilities do not merely constitute a "last line of defence" when other barriers fail or are missing. They are the ones that make the other two barriers effective. Indeed, if we do not develop our human factors and orient them towards safety as part of a holistic prevention, protection, and reaction system, everything else just falls apart.

The everyday experience reminds us that a technological barrier that is perceived as a constraint by the workers will not be truly effective unless they accept such a constraint. We have stopped counting the number of accidents derived from situations where the operator cheated a safety interlock switch, removed a piece of guarding, or deactivated an engineered safety feature because it imposed a supplementary effort on them or made the task longer to complete.

And what about organizational controls? Rules, procedures, SOPs... they are worthless if the men and women who must abide by them are not intimately convinced of their purpose and committed to safety. If that is not the case, the very human nature that we seek to positivize will push them to the path of least effort.

We can see that human factors are not just the last barrier but also, and chiefly, the *glue* that holds together the whole structure of prevention, protection, and reaction control means (Figure 4).

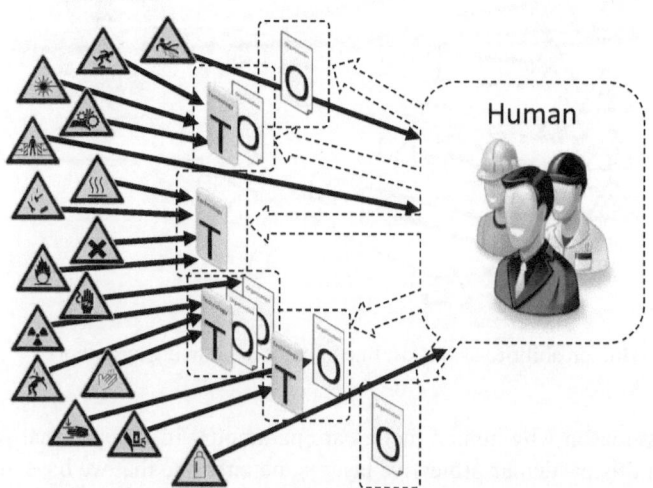

FIGURE 4   Humans as a transversal factor (drawing by author).

BEHAVIOURS AND OCCUPATIONAL HAZARDS

"Precaution is better than cure".

Johann Wolfgang von Goethe

By contemplating human factors – both at individual and collective levels – as the key to occupational risk mitigation, this book intends to look for the root causes of safety rather than those of accidents. We will unveil the success factors behind the brilliant safety track records of best-in-class organizations, because it is not by sheer luck that one attains and sustains excellent results in this domain.

Turning the argument on its head opens our minds to *looking at safety in a positive way*. And our organizations sorely need this after decades of defining it in negative terms. Intuitively, most people consider safety as the *absence of events* causing damage. Consequently, the most obvious way of measuring safety in the workplace is to just count the number of accidents and assess the severity of their consequences.

Most organizations, from agencies in charge of nuclear operations to large corporations, as well as small businesses, categorize and account for accidents and incidents to calculate frequency and severity rates. This is definitely important information, but very much like a thermometer that tells that you have a fever, it does not give you any indication on what to do to heal. We will come back to the slippery topic of measuring safety; for the moment, let us examine the concept of safety a bit more closely to schematically draw the plan of the next chapters.

About Hazards and Risks

Most people are not intrigued by safety out of mere academic curiosity but out of a desire to prevent damage and losses, foremost loss of life, and the most effective way to achieve that is to avoid accidents. This requires identifying the hazards and assessing the risks linked to a given operation (Figure 5).

The evaluation of the potential severity of a possible accident and the likelihood of its occurrence (in general related to the frequency of exposure to the hazardous element or situation) determine the risks' intensity. This reckoning, whether deliberate or not, will conduce to the appreciation of the risks' acceptability with regard to the risk appetite (or aversion) of the individual or organization. The archetypical graphical representation of the combination of these two factors is a double-entry matrix with the cases in the top right corner coloured in red, those in the bottom left corner coloured in green, and those in the middle diagonal in different shades of yellow and/ or orange.

By identifying hazards and assessing the associated risks, a necessary first step is taken, but it does not yet get us out of the negative concept of safety. So let us take a bit more perspective. What will determine the nature of the hazards present

FIGURE 5   From hazard to damage (drawing by author).

in a given workplace, either when operators execute a very particular task or as a result of the interaction of multiple and diverse activities? What defines the type and frequency of the workers' exposure to those hazards, as well as the kind of control means that are deployed and their efficacy?

FOCUS ON... ACTIVE FAILURES, LATENT CONDITIONS, AND MURPHY'S LAW

In the late 1980s, James T. Reason proposed his classic "Swiss cheese model" of accident causation, which aims to illustrate that, for a hazard to cause an accident, both **active failures** and **latent conditions** must combine.

Active failures are typically unsafe acts and execution errors. Latent conditions, however, are flaws in the system that may be dormant for a long time. Reason compared the latter ones to "resident pathogens" that, unnoticed or tolerated due to compensation by other system features, remain hidden until they align with one or more active failures, thus creating the conditions of the accident.

Humans being human, and hence fallible, which tends to be problematic when coping with danger, Reason pointed out the limits of the safety approaches that focus on the operator and which try to just prevent unsafe acts (and notably behavioural safety, which we will discuss later) and advocated for the *systemic approaches.*

Such systems must account for the possibility of active failures, unsafe acts, and errors by the operator, and be able to deal with their consequences. Simultaneously, the system must ensure that the necessary processes are deployed to identify and correct its own flaws, the latent conditions resulting from decisions, missteps, and errors that took place upstream, often far apart in time and space from the frontline (Reason, 1990, 2000).

Stricto sensu, preventive action can only target latent conditions, which can be spotted and eradicated. Since active component failures can take all kinds of shapes and forms and happen anytime, they shall be considered unforeseeable in their timing and particularities but predictable in absolute terms, because they are unavoidable in the long run: anything that can go wrong *will* go wrong. As Edward A. Murphy Jr. put it in his original quote, "if there are two or more ways to do something and one of those results in a catastrophe, then someone will do it that way" (Bloch, 1978). Using Reason's logic and jargon, the system allowing for a way of doing something that results in a catastrophe is a typical latent condition that must be chased. But fallible as we are, we will not be able to detect and suppress all of those. A practical approach should consist of making sure that we can cope with both active failures and latent conditions by developing the system's *resilience,* a concept that we will explore later.

OPERATIONS: ENGINEERING AND EXECUTION

The dangers that workers will face when doing their jobs are determined by the way the operations are executed in each instance, both individually and collectively – as well as by the protective barriers in place. We also know that these factors can be very different depending on the context, the moment, and the person.

So, we have human beings at operational, support, engineering, design, management, and executive levels making choices, taking decisions, and performing all types of tasks. In short, we are confronted with the **behaviours** of many different players interacting with and influenced by each other, their physical and social environments, and a multitude of targets and constraints.

Behaviours are what people do and what they do not do. For over a century, scientists have been studying this completely observable phenomenon, yet its roots remain hidden, rendering its management excessively complicated. Behind this simple concept, we are mixing things as different and elusive as decisions and choices, automatisms and habits, and actions and reactions – some of them carefully reflected, some rather intuitive and some totally spontaneous. To summarize, *human factors* are at work everywhere in our organizational and cultural environments.

Before going any further, it is very important to insist on the fact that we will fully consider the upstream behaviours, those that occur at executive and management levels in every organization. When we hear about behaviours and human factors, more often than not, it is the action of the front-line worker that is targeted and questioned. We are quick to point fingers at the error, the deviation, the violation, or the "silly mistake" of the one whose hand dropped the ball. The one who brought the final straw that broke the camel's back.

It is paramount to bear in mind the choices and decisions, individual and collective, of all the players involved in designing the processes and the working environment, and, above all, those made by executives, decision-makers, and people holding the authority to set profitability targets, allocate resources, and grant or deny budgets. This is because, either directly or indirectly, they will determine the material, organizational, and even moral conditions on which operations will be performed and the job will be done (Figure 6).

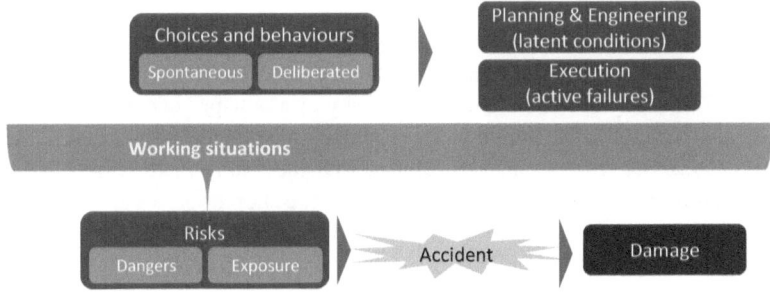

FIGURE 6   From behaviour to damage (drawing by author).

FOCUS ON... THE SAFETY EQUATION

Visions and slogans like "Zero accident" or "Zero harm" have the merit of rallying the teams behind a clear-cut and simple message. They are, of course, commendable as the ultimate goal (in any case, how many accidents would you suggest as a target for your organization?). However, we must acknowledge that they are not realistic in the occupational health and safety arena. In the workplace, there will always be circumstances that cannot be controlled or even foreseen. This is why we propose the following equation as a simple way to visualize the safety level of an operation:

$$Safety = Behaviours \times Luck$$

The probability of having our workforce returning home safe and sound after work depends on these two variables:

• The behaviours of every player, from the front-line operator to the final decision-maker whose saying may have something to do with the operating conditions, and
• How lucky they will be that day, chance being, by definition, the result of everything that they cannot control with the means at their disposal (financial, technical, human...).

Empirically we observe that these two variables are negatively correlated, acting as communicating vessels: the better we succeed in orienting the behaviour of every player towards safety, the less room we leave to "bad luck". This proposition intends to find a middle ground between *chimerical determinism,* making us believe that every single accident can and should be avoided, and *defeatist fatalism,* telling us that whatever we do, there will always be too many accidents.

WHAT WE DO

Legendary basketball coach and player John Wooden once said, "it's not what you do, it's how you do it". With all due respect, we can argue that what you do certainly matters too. We may imagine a new technology or a new project and decide that the level of risk is acceptable. We conceive and design workplaces, processes, and barriers to protect the workers and bystanders that inevitably include flaws and blind spots, which Reason called latent conditions. On the shop floor, we grasp, drop, push, pull, step, click, and execute thousands of other motions and gestures, including, inevitably, the occasional blunder – under different shapes and forms – that he called active failures. When the former meet the latter, we have an accident. Our physical and mental abilities allow us to perform optimally most of the time. Yet, we are bound to reach the limit of these capabilities at some point because, yes, we are fallible, and so are the systems that we conceive and build.

To revert back to our positive vision of safety, we shall remember what Weick said: although "reliability in invisible and [...] operators see nothing and seeing nothing presume that nothing is happening [...] this diagnosis is deceptive and misleading because dynamic inputs create stable outcomes" (Reason, 1997). Our proactive and adaptive behaviours – Weick's "dynamic inputs" – ensure our safety much more often than they cause accidents.

Behaviours and our inherent imperfections have been traditionally studied from philosophical and moral perspectives. In every civilisation, religions and schools of thought have offered explanations for our ways of behaving and proposed precepts to guide them to avoid harm. Of course, these are very much present today around the world – and we will see that our new understanding of the role of emotions is giving them a newly acquired scientific legitimacy – but the behaviour phenomenon did not escape the 18th century's scientific revolution, resulting in the blooming of economics, psychology, and so many other human and social sciences in the 19th century.

The theories elaborated by these sciences shaped the understanding of judgement, choice, and behaviours that prevailed during the 20th century and that continue to impregnate our current paradigms. Originally, none of those disciplines cared much about occupational health and safety, but these ideas have fashioned the role that was assigned to the workers in the risk mitigation strategies that were deployed since then. Fortunately, science has progressed, theories have evolved, and so have ideas in the occupational health and safety field.

In the following chapters, we will discuss the most relevant of those theories, old and new, to reveal the nature of our behaviours at work, both at individual and collective levels. We need to know their fabric inside out if we want to be able to orient them effectively and sustainably.

The first series of phenomena that influence an individual's behaviour in a given context is the mental representation that they construct of the situation. Generated on the basis of our **perception** of reality, this representation integrates the data we collect from the situation and our preexisting mental models, including the representation that we make of ourselves and our role in the situation. In Chapter 2, we will uncover human perception's capacities and limits, both in terms of sensors (our senses) and signal treatment (our brain), which prefigure our subsequent cognitive and affective processes.

With this first step, we will introduce the analysis of the different compartments of our **memory** that we will pursue to better understand key phenomena like our **situational awareness, attention,** and **concentration**. If we ignore the fundamentals of these abilities, their strengths and weaknesses, we may very well end up implementing inadequate risk management strategies and techniques. Some may expect every worker to be able to catch and decipher, within a split second, all the signals required to assess the risks they are exposed to. This is not a given, and we need to recognize the complexity of these processes to address them properly and adapt the job to the worker.

We will see that even at these early stages of our cognitive processes, the individual's **values** and **beliefs** already play a fundamental role. We will take a close look at these keystone elements that indeed influence our perception – as they affect higher cognitive functions. Concurrently, it is the perception of our reality that shapes over the long term our values and beliefs in a feedback loop that we will dissect.

As we will explain in Chapter 3, both science and our quest for practical methods of influencing behaviours at work suggest that there is a lot of merit in considering safety (a term that shall be considered as a synonym for both prevention and occupational health and safety in these pages) as a value on its own. Safety is a value that is linked in our psyche to a set of specific beliefs and to other values. Conversely, it can also be confronted with competing values in a process that ultimately shapes our **attitudes**.

We will present a wealth of research that shows how, once they are established, our attitudes durably influence **reasoning**, as well as **emotion** and **motivation**, via a series of mechanisms that we will examine. We will uncover how such mechanisms ultimately trigger and control our behaviours at both conscious and subconscious levels. Attitudes can be considered as their common roots, but there are, of course, major differences between deliberated behaviours and automatic ones, and we shall examine them in detail.

We tend to see ourselves as fully rational entities, but we need to realize that we are very often prisoners of our **reflexes**, **habits,** and **routines**. We will expose the situations and conditions when these serve us well and when they may lead us to harm. Moreover, we will see that even when we are prompted to reflect and "think", our **rationality** is not always as robust and reliable as we think it is. The latest discoveries in neuroscience and experimental psychology are shedding a very sobering light on these aspects (Figure 7).

Once we have a solid grasp of these phenomena at the individual level, we will be in a position to look at them in their organizational context, considering their social and cultural implications. Because it is mostly at this collective level that we must act in the frame of occupational settings. In Chapter 4 we will describe the ingredients

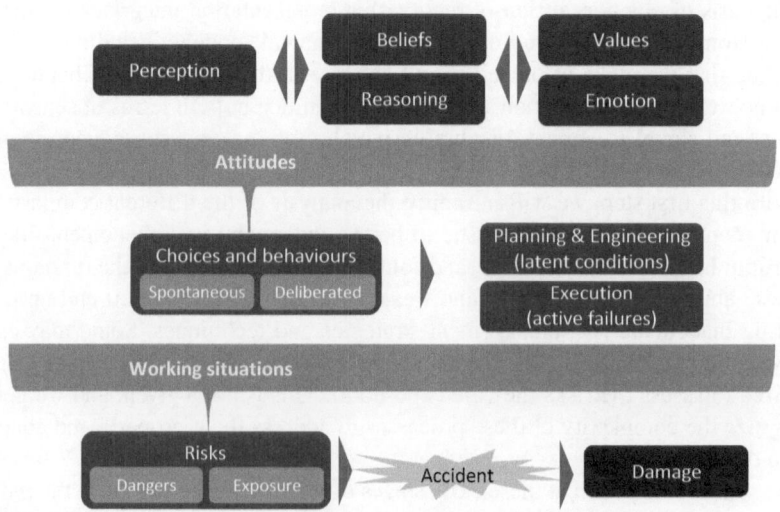

FIGURE 7   Multifactorial origins of behaviours (drawing by author).

of what we now call **organizational culture**, how they interact with each other, and how to address them so as to make the culture evolve in the desired direction.

To that end, we will finally review the most relevant of the many different approaches to **safety leadership** and how, independently of styles and idiosyncrasies, there is a set of qualities, techniques, and tools that can help us promote occupational health and safety and place it at the very heart of the organization's culture.

All the concepts that have been briefly introduced in these lines interact with each other within highly sophisticated sociotechnical systems. In Chapter 5 we will summarize these interactions in a synthetic model that aims to provide the reader with a holistic vision of human behaviour in the collective occupational settings and the main keys to orient it towards safety.

Since we must present them as a sequence, we will start our journey with those that are the subject of Natural Sciences. We will later move on to those that are studied by the Social Sciences.

2 The Brain
Perception, Attention, and Memory

"Reality is merely an illusion, albeit a very persistent one."

Albert Einstein

Understanding human factors requires us to understand the functioning of our senses and our brains. Considered as the most complex organization of matter in the universe, the human brain is a fascinating organ that we are still discovering. It is composed of some 100 billion neurons that form a dynamic network of over 100 trillion synaptic connections.

DID YOU KNOW? COMPUTERS HAVE MATCHED OUR BRAIN'S COMPUTING POWER, BUT NOT ITS EFFICIENCY

The order of magnitude of our brain's neural computing power is the exaflop (a billion billions operations per second). This figure was reached in June 2022 by a supercomputer called Frontier. It required 22.7 MW of electrical power to accomplish that feature, whereas our brain is a million times more frugal, requiring just between 20 and 40 W (the equivalent of a small incandescent light bulb).

Incidentally, our brain must fit in a skull whose volume needs to remain compatible with pelvic delivery, while Frontier is composed of 74 19-inch (48 cm) rack cabinets occupying 7,300 square feet (680 m²).

This stunning biological "black box" within our heads is structured in dozens of hyper-sophisticated interconnected neural structures. They are the core of our nervous system, which is itself connected to our endocrine system and to the rest of the body by sensory and motor nerves. Electrical impulses and chemical synapses between neurons allow information to flow and support cognitive functions such as perception, attention, concentration, and deliberation based on different memory systems. These functions, in turn, enable us to think, imagine, feel, and be moved – the faculties that make us human, for better or worse.

Even though the power of our brain has made humans the dominant species on this planet, it comes with some specific features and limitations that we tend to ignore. As we will see, knowing them better can save us from some potentially fatal pitfalls.

DOI: 10.1201/9781003617341-2

PERCEPTION

"It's not what you look at that matters, it's what you see".

Henry David Thoreau

Before reaching our brain, information acquisition starts with our senses capturing external and internal *stimuli*. We name in this way the physical, chemical, or biological elements that can trigger reactions in our bodies, notably in the nervous, muscular, and endocrine systems. Signals from both within and outside our bodies are translated into electrical and/or chemical messages, which are addressed to the brain and more precisely to the sensory cortex. More precisely, to the sensory cortices, as we possess several of those, the main ones being the visual, auditory, olfactory, and somatosensory cortices.

The sensory cortices hold our **perceptual memory**, an implicit, nonconscious, effortless, and automatic memory that can be immediately used by our cognitive processes and can also be preserved there for years. The prefrontal cortex plays a role in integrating real-time sensory information with prior sensory data (*percepts*), thus stabilizing our sensory experience (Schwiedrzik et al., 2018).

Signal Acquisition

Let us first take a closer look at the purely physiological aspects of stimuli capture, as they impose a first set of limits to our perception. Among our senses, our human species has developed one that is our primary source of information from our environment to the point that the others could almost be considered "secondary". We will start with those and leave the most relevant one for the end.

Secondary Senses

Our sense of smell can be counted among the senses that bring relatively little information to us. However, this does not mean it is not useful, sometimes to the point of saving our lives: our **olfaction** enables us to detect and recognize airborne chemical substances, whether in the form of gas, droplets, or dust. This is highly convenient when it comes to taking notice of a fire start, as well as the presence of organic volatile compounds that can be potentially dangerous for our health and even flammable.

Different individuals can have a more or less developed sense of smell, and it can be "trained" with practice. And like our other senses, olfaction can be exacerbated by our attention, and the perception of smells can be modulated by subjective and cultural factors that we will study later.

However, the fact remains that human olfaction is not very efficient: the dog's sense of smell is 1 million times better. We can enhance our olfactory capabilities, but we can also lose them: one can suffer from *hyposmia* (decreased sensitivity to one or more smells) that can be temporary or permanent, and even from *anosmia* (total loss, aka smell blindness). These conditions may be congenital or acquired, caused by head trauma, drugs, neurodegenerative diseases, and upper respiratory infections – they became widely known as both symptoms and sequelae of COVID-19.

Furthermore, the olfaction receptors tend to saturate rather quickly, so overall, the sense of smell cannot be considered either very preeminent or highly reliable with regard to our perception.

DID YOU KNOW? YOUR ABILITY TO DETECT SOME DANGERS DEPENDS ON YOUR MUCUS

It is the olfactory mucosa that allows us to capture the scents carried in the air. It represents a surface of barely $2\,cm^2$ (less than $1\,in^2$) located inside our nasal cavity, which is lined with olfactory epithelium where terminate some 5 million olfactory neurons. To compare once again with our favourite pets, dogs' olfactory mucosa possesses from 100 to 300 million, depending on the race.

Approximately 400 types of olfactory receptors have been listed, each one based on a particular protein that recognizes a given molecular structure, hence having a specific sensitivity. Some molecules can be detected at much lower concentrations than others, like the thiols (organic compounds containing a sulfanyl group, also known as mercaptans) that are used as odorants to help people detect leaks of natural gas, which is odourless.

Once a receptor has caught a matching molecule and the information is sent up the olfactory neuron, it will not resend the message. The olfactory epithelium must be continuously flushed by the renewal of its mucus. This mucus not only absorbs molecules from the air and brings them into contact with the receptors but also cleans them; otherwise, they saturate. This is why you stop noticing the scent of the flowers in the room after a moment.

Blowing your nose rids your olfactory mucosa of soiled mucus loaded with germs and is a healthy habit that prevents infections and respiratory disease. It also "resets" your saturated olfactory epithelium. Sneezing is an innate reflex that naturally achieves those same goals.

Hearing is another sense that helps us collect information that can be vital for our safety. Unfortunately, many working environments are too noisy. The threshold of 85 dB is the most popular amongst regulators around the world for mandatory hearing protection to preserve the workers' hearing. But noise levels significantly lower than that can mask sounds that may otherwise reveal an issue, hazard warnings, or safety instructions. Earmuffs and earplugs used for hearing protection can have the same masking effect, protecting the user's health but potentially depriving them of valuable information. State-of-the-art hearing protection devices now integrate passive and/or active filters that block the noise while letting human voice's wavelengths pass.

Communication and coordination between team members can also be affected by noise, sometimes in subtle ways like hindered processing of social signals, and some other times in very flagrant manners like heightened emotional reactivity and aggressiveness. Because noise can be disturbing, it produces distress and anxiety, affects concentration, reduces awareness, and even negatively influences brain performance and decision-making. A correlation between occupational injuries due to "unsafe behaviour" and noise exposure has been found in a study carried out in Indonesian construction sites, suggesting that noise as low as 75 dBA results in psychological changes (Prayogo et al., 2024).

Furthermore, our hearing naturally deteriorates over time. Aging, as well as long-term exposure to noise, inside and outside the workplace, leads to hearing loss. Similarly to our olfaction – although the process is quite different in its details – our hearing also "saturates" when exposed to constant sounds. These are then physiologically filtered out and will not reach the brain for processing.

DID YOU KNOW? THE FITNESS OF A SNAIL IN YOUR EAR CAN AFFECT YOUR HEARING

In our inner ear lies a small spiral-shaped organ called the **cochlea**, the name derived from *kokhlias* (snail shell in Ancient Greek). It is divided into several sections, three of which are filled with fluid. The cochlea magnifies the pressure of acoustic waves by reducing by 20 the area ratio from the tympanic membrane to the oval window, resulting in a 20-fold pressure increase.

The amplified vibrations in the fluid cause the hair cells present along the entire cochlear coil to move. Hair cells are a special kind of neuron that generates action potentials, which are relayed by other nerve cells up to the brain, but some processing occurs in the cochlea itself. The cochlea not only amplifies sound by the reduction of the area ratio mentioned above, but it also does that by *generating* vibrations. The outer hair cells can convert electrical signals back to movement, causing vibrations that will be detected down the coil by the inner hair cells, in a positive feedback loop. This is a key feature to capturing weak sounds that also explains a first "noise reduction" mechanism, as this loop can be unconsciously deactivated when we deem that the sounds perceived are irrelevant.

Hair cells do not regenerate when damaged, whether by age or exposure to excessively loud noise. Outer hair cells are more susceptible to damage and death, resulting in diminished sensitivity to weak sounds.

Like olfactory receptors are specific to particular molecular structures that are then processed as different scents, hair cells are tuned to certain frequencies that are processed as different sounds. Interestingly, the cochlear response to the volume of sound is not linear, as some tones reduce the neural firing rate generated by a second tone, a phenomenon known as two-tone rate suppression.

We could also speak about the sense of touch or about proprioception (our ability to pick up stimuli coming from our own body, not to be confused with interoception, which we will study later).

Vision

But among our senses, it is our **sight** that brings us the most information about our environment: 80% of the stimuli processed by a healthy person's brain are visual.

Consequently, almost a third of our brain is dedicated to the processing of visual information (Wandell et al., 2007).

DID YOU KNOW? VISUAL CORTEX RECONVERSION

People born blind and those who lose their sight at an early age tend to develop their other senses extraordinarily well. Their enhanced sense of touch enables them to read Braille, and their acute hearing allows them to localize the source of sounds. Some blind people have extended this echolocation ability to the point of being able to locate walls and other large obstacles while walking by emitting sounds, generally clicks. They can interpret as spatial information the echo that comes back to them when the sound bounces back.

This is possible thanks to our brain's **neuroplasticity**. Our neurons can evolve and reorganize, so sections of our visual cortices can be dedicated to other functions. We will explore this phenomenon further, as it plays a major role in other cognitive processes.

For individuals enjoying fair visual acuity, the prevalence of sight is such that it influences our brain's interpretation of stimuli coming from other senses. Maybe you have already been on a train waiting to leave the station and had the impression that it started moving, whereas it was the train on the platform next to yours that had got started. Similarly, individuals tasting candy or jelly of different colours can appreciate different flavours – lemon if they are yellow, strawberry if they are red – even when the only difference between them is the addition of a tasteless colourant.

On a more dangerous note, we can subconsciously ignore other stimuli, for instance, sounds, which would appear to be in contradiction with what we see. Beeps from reversing radars may not be noticed by an automobile driver who does not see the obstacle in the rearview mirror. Pedestrians and workers in the proximity of trucks and construction machinery may not effectively perceive the sound of warning audible reverse devices if the vehicle is not in their line of sight.

FOCUS ON... THE MCGURK EFFECT

In the 1970s, psychologists Harry McGurk and John MacDonald were studying speech perception in children. When they inadvertently presented to them video footage with a soundtrack out of sync with the image, they realized something strange. If the sound "ba" was emitted but the image showed someone saying "ga", it was impossible for the subjects to recognize the sound "ba" that they were actually listening to. Most of them declared hearing a sort of "da" or "ta" (McGurk, 1988).

The fact of seeing a person articulate a sound without their lips making contact with each other makes our brains refute the sound "ba". Our life experience tells us that to pronounce the sound "ba", lips must touch. This is an excellent illustration of the influence of our learned mental models on our perception, an influence that also affects other cognitive processes that we will examine.

So, we rely mainly on our vision, a system that usually serves us well. In fact, it works so well most of the time that it can be problematic when we are confronted with its limits. We are so used to getting an accurate picture of what we see that we can be tricked.

The Limits of Our Vision

The first limitation relates to the individual's **visual acuity**. We are not all born equal in this domain, and even when our vision at birth is alright, it can deteriorate over time. The prevalence of visual impairments in the population is growing fast, especially in the developed world. The ever-increasing number of hours that we spend watching closely – traditionally at texts and books, and now looking at screens – and the diminishing amount of time that we spend outdoors, gazing into the distance, is taking a toll on our visual health.

The individual's vision most often degrades progressively over the years. Our brain grows used to it but is unable to compensate for the increasing amount of missing information. A more or less blurred vision becomes normal to us, causing many people to underestimate the importance of their vision's deficiencies.

DID YOU KNOW? MOST PEOPLE HAVE SIGHT PROBLEMS

In 2000, almost 30% of the world population suffered from *myopia* (shortsightedness), a figure that is projected to reach almost 50% by 2050 (Holden et al., 2015), while *hyperopia* (farsightedness, aka hypermetropia) affected 10% of the population in the United States (Trobe, 2006). Around 70% of people aged 40 and more suffer from *presbyopia* (Fricke et al., 2018).

Globally, over 2.2 billion people have a near or distance vision impairment, with the leading causes being refractive problems listed above and *cataracts*. It is estimated that only 36% of people affected by refractive error and 17% of those affected by cataracts have access to appropriate care (WHO, 2024) and that 23% of drivers have such vision impairments but remain uncorrected (WHO, 2020).

Another form of impairment is colour vision deficiency (aka colour blindness) related to the functionality of the cone cells in the retina. The most common type, congenital red–green colour deficiency, affects 8% of men (Birch, 2012). They have decreased or no colour discrimination along the red–green axis. As colour-coding is broadly used in workplaces and signalling, especially red and green, red–green colour deficiency can be problematic for safety. Traffic lights on the roads, railway signal lights, and navigation lights in marine and aviation settings are based on red–green–yellow colours. These colours are also used to identify cables in electrical circuits, and more broadly, the red colour is largely used to identify danger, access restrictions and fire-fighting equipment.

Why do we use the red colour so extensively to draw attention, one may ask? Aside from colour-blind people, almost two-thirds of cones process the longer light wavelengths, which have better penetration power (not only reds but also oranges and yellows) and are less scattered as they travel through the air towards our eye. Being the colour of blood, evolution and natural selection probably favoured the development of such receptors, allowing us to detect injuries in ourselves and our tribe members and track wounded animals when hunting.

Even if we do not suffer from any particular visual pathology, there are some more or less obvious limits to our vision.

It may seem unnecessary to point it out, but, at best, we only have a pair of eyes, and both of them are in our face, oriented forward. This prefigures the width of our **visual field** and necessarily means that there is a lot of information that they just cannot pick up, hidden in what we could call *anatomical blind spots* (Figure 8).

FIGURE 8   Drivers' blind spots (artist unknown).

Our visual field is limited horizontally to some 200°–220°: around 110°–120° for binocular vision (required for depth perception, aka 3D vision) plus circa 50° on each side that can only be accessed by one eye. However, it is also limited to a vertical range of around 150°. Of course, this does not mean that we can see everything in that range, as many objects will be simply masked by obstacles.

It is troubling how often we ignore this simple fact and operate under the impression that we have access to all the information we need to move around, whether as pedestrians or when driving a vehicle. We will get to explain how this happens in the following chapters.

Maybe less evident, the very architecture of our eye and the way its photoreceptors are placed in our retina impose limitations on the completeness of our vision.

When we look, the visual field described above is divided into two different sections:

- The **central vision** (aka *macular* or *foveal vision*), which is ensured by the central part of the retina. In this small area called the macula lies an even smaller, central pit called the fovea, where most of the *cone cells* mentioned above are located. These cones are the ones that perceive colour, finer detail, and rapid changes in the images but are not very sensitive in lower light levels. The fovea is so small that this central vision represents barely 2° of our visual field. Interestingly, when in dark conditions, it is possible to not see a bright spot that you would be staring at directly, as long as it is caught by our central vision only.

FIGURE 9   Central vs Peripheral vision (drawing by author).

- The **peripheral vision** (aka *indirect vision*), which embraces everything else outside the point of fixation of our central vision. It is performed by the larger, outer part of the retina beyond the macula, where we find essentially *rod cells*. More numerous than the cones (we have almost 100 million rods vs some 6 million cones), the rods are complementary to the cones in their function: they are about 100 times more sensitive to a single photon than cones, and we owe them our night vision. In addition, multiple rods converge to amplify the light signals. But this convergence pools information and thus reduces the image resolution. Furthermore, rods do not distinguish colours.

This means that when reading this book by holding it at about 40 cm (1.3 ft) from your eyes, you only get to see clearly a surface of roughly 5 cm^2 (0.08 in^2) of the page you are looking at, the equivalent of a 2 euro coin (less than the surface of a quarter dollar coin).

On top of that, there is a part of our retina called the **optic disc** (aka *optic nerve head*) where the optic nerve begins, carrying over 1 million afferent nerve fibres from the eye towards the brain. As there are absolutely no cones or rods in the optic disc's area, it represents a small *physiological blind spot* in each eye.

How come that we do not realize these limitations, the incompleteness of our sight? In fact, our brain compensates for these structural constraints; it plays a smart trick and keeps our eyes scanning our visual field. Unconsciously they perform some 100,000 *saccades* (French for "jerk, bump") every day, orienting our fovea to explore the details of the image in front of our eyes. Our perceptual memory integrates all these inputs into a seemingly seamless image, but our visual perception is a construct put together on the basis of fragmented information (Figure 10).

FIGURE 10   The skull and the dot (drawing by author).

TEST NO. 1: The skull and the dot

Carefully follow the following instructions:

1. Take this page with your left hand and hold it at arm's length [if reading the e-book version, just place your face at arm's length from the screen].
2. With your right hand, cover completely your right eye.
3. Lock your sight on the black dot. Remember, your brain will unconsciously push your open eye to scan your field of vision (the *saccades* mentioned above): make a deliberate effort to keep it locked on the black dot.
4. Very slowly, bring the book towards your face, while keeping your open left eye locked on the black dot [if reading the e-book version, very slowly get your face closer to the screen].
5. Somewhere midway, something funny should happen…

If you thoroughly follow the instructions, the skull just disappears. This experience is devised to put it precisely on the *physiological blind spot* caused by the optic disc in your retina, described in the previous section. Normally we do not have the opportunity to realize this phenomenon, first documented by French physicist Edme Mariotte in the 17th century, without playing a purpose-designed game like this one.

But what is most interesting about this experiment is that it reveals what our brain does when facing missing information: it just "fills the gap" with what seems most likely to be there. As the close environment to the blind spot is just white, empty space, our brain subconsciously infers that most probably there is the same in that spot for which it lacks information.

We could hope for some kind of unease or other warning sign telling us that we are missing what may be important information for our safety (the skull shape is meant to convey a message of danger, after all). But our brain is programmed to behave in the completely opposite way: it will reassure us by producing a seamless picture, but one that is inaccurate and could put us in harm's way.

To crown it all off, the working environment does not always help us:

- **Lighting conditions** are not always optimal, especially in non-routine operations. Truck drivers, maintenance technicians, and construction workers, to give just some examples, often work in environments that can be quite dark, but they can also be dazzled by ill-positioned light spots. Glare can momentarily blind us, and both in the dark and in the glare, we are unable to accurately recognize colours.

FIGURE 11   Glare and driving (picture by author).

- **Particles** suspended in the air (or water if in subaquatic works), like dust, fog, and smoke, interact with the light and absorb it, especially the shortest and most powerful wavelengths (violet and blue) as mentioned above. Therefore, fumes, smog, and mists are not only dangerous for your health, but they can also prevent you from acquiring the visual information you need to perceive dangers and work safely.
- **Eye protection** such as goggles, safety glasses, and other kinds of masks is, of course, necessary in many instances, but it can hinder the workers' vision. Their lenses or screens tend to be quickly scratched, and usually, workers do not cleanse them as often and as thoroughly as they should.

What is truly fascinating about all these limiting factors, whether anatomical, pathological, physiological, or environmental, is that they go unnoticed by most people most of the time. Habituation, a psychological phenomenon that we will study later, plays an important role here. How many times do we realize that our spectacles were dirty only when we take them off?

We sincerely believe that our senses, and our sight in particular, provide us with an accurate image of our working environment. We operate under the impression of having acquired all the information that we need. We are surprisingly at ease most of the time, considering all the facts that we are missing every single moment. And this ease may very well lead us to neglect some verifications that can seem superfluous to us if we believe that we perceive everything.

Key Learning Points for the Safety Leader

The raw data that reach our brain, on the basis of which we construct our perception of reality, are far less complete than we use to believe.

In our species, these raw data consist chiefly of visual stimuli, and our brain will give priority to visual information, to the point of subconsciously ignoring contradictory stimuli (for example warning sounds that are not confirmed by our sight).

Even if we are among the minority that is fortunate enough to have perfect sight, our vision is much less accurate than we think. It is the image processing by our brain that allows us to make the most out of the fragmented and incomplete set of data that it receives, as we will see in the following section.

TOOLBOX – Tool No. 1: An Additional Third Point of Contact

The rule of three points of contact is today an absolute classic for avoiding falls when climbing up and down ladders and stairs. It has been historically used in offshore oil & gas and marine operations and is today widely applied in industrial settings as well as in transportation.

The two basic points of contact (your feet, which, as long as you are not running, should both touch the floor) are complemented by the fact of holding at least one handrail, eventually two in very step ladders.

Unfortunately, this interesting concept cannot be applied in the absence of a handrail... or could it?

The author has been advocating for years now for the adoption of an "additional third point of contact" that can help us not only avoid trips and falls in ladders and stairs, but everywhere.

We can consider our attentive and deliberate visual contact with the step, as well as the floor that we are about to step on, as being an effective, although "hand-free", point of contact.

This third, visual point of contact not only effectively helps us to avoid the pitfalls related to missing an obstacle in our path, as noted when we presented the limits of our visual field. It also helps us keep our balance and successfully orient our motricity, which is very much directed by our gaze, as we will discuss in the section dedicated to our sensorimotor skills.

Training shopfloor personnel to mindfully use their gaze as a point of contact with their path when moving around, whether on foot or when driving a vehicle, is an advanced tool. It may seem ludicrous to workers and managers that are not yet fully familiar with human factors. A preliminary awareness of the workforce on the functioning of our brain appears necessary before proposing this tool.

DATA PROCESSING

Whereas the signals acquired by olfactory neurons are transferred rather directly to the olfactory bulb in the brain's prefrontal cortex, hearing and visual perception start in the cochlea and retina, respectively. After this first processing, signals are sent in the form of electrochemical energy via the thalamus towards the primary auditory cortex for the former and to the primary visual cortex for the latter.

There the heaviest processing work begins. Neuroscientists have unveiled the role of several different parts of our brain in the processing of different stimuli, but perception requires many areas to work in a coordinated manner. It is not a linear process but rather an emergent property of an incredibly complex system.

FOCUS ON... FROM PRÄGNANZ TO PREDICTIVE CODING THEORY

Our **perceptual memory** integrates the stimuli coming from our senses, but it shall not be considered as a mere repository collecting data coming one-way from the rest of our body.

Back in 1867, after examining the human eye, Hermann von Helmholtz reached the conclusion that it was not capable of producing high-definition vision as we experience it. He called *unconscious inference* the process through which our brain makes assumptions and constructs a complete visual impression from incomplete data, based on our previous experiences (Von Helmholtz, 1925).

The *Gestalt* (German for "form") theory on perception resumed this postulate at the end of the 19th century when it affirmed that, rather than collecting and adding data representing independent details, we mostly perceive wholes that are processed summatively. The fundamental principle of this perceptual grouping is the law of good form or *Prägnanz* (German for "pithiness"). Our life experiences make us perceive the elements that we see as wholes forming simple and familiar forms.

Our brain does this grouping by using characteristics such as proximity, similarity, continuity, and closure (Köhler, 1947). In Figure 12, continuity and closure lead our brains to follow the *reification* principle and see a cube: the experienced object of perception contains more information (a 3D object) than the sensory data on which it is based (eight stripped circles).

Afterwards, the *predictive coding theory* and the *Bayesian Brain Hypothesis* have suggested that our brain operates following a predictive model that generates sensory expectations (Wacongne et al., 2012). According to these, the previous image evokes a cube chiefly because it is a very familiar shape that we see often.

This predictive ability would allow for a simplified data architecture and processing: only the stimuli that do not correspond to our brain's prediction would be singled out for further analysis. It would also enable us to interpret noisy data and even fill the gaps left by missing data.

FIGURE 12 The imaginary cube (drawing by author).

TEST NO. 2: The Thatcher effect

Do you recognize the person in the picture below? Normally we do not have much trouble, since she was a public figure, and we have seen her face many times.

Try now and turn the page upside down. Most probably you had not realized to what point the image was distorted! Both her eyes and mouth have been tampered with and, unlike the rest of the face, were in the right sense.

We call this trick the Thatcher effect because it was first demonstrated in 1980 in a photograph of the then British Prime Minister Margaret Thatcher by Peter Thompson, professor of psychology at the University of York (Thompson, 1980) (Figure 13).

This phenomenon has been used in the study of the psychology of face recognition, revealing the predictive and holistic nature of the processing of facial images. We do not examine their separate elements (nose, eyes, mouth…) but rather process them as a whole.

This same predictive and holistic processing makes it so that we can very easily miss particular details when confronted with familiar objects (like machinery, control panels, and working tools) and environments (like our

FIGURE 13   The Thatcher effect. (From Peter Thompson (1980)).

workstation). We give a general glance at them, and if everything "looks normal", we just go ahead, possibly missing more or less subtle anomalies and signs of danger.

Meaning and Sense

Another very important takeaway from these discoveries is that we will probably not perceive something (see it, hear it, smell it, or feel it) unless that something has a meaning for us or we are familiar with it from previous experience. We may notice the parts, the elements, that there is "something there", but we may fail to perceive the thing itself. In Goethe's words, we only see what we know.

And even if we know the object at hand, if we fail to recognize it, we will most probably fail to perceive it as well. As we realize that recognition and eventual perception are related to our expectations, we can understand that we often *miss objects that we do not expect to be there*. This is a major contributor to slips, trips, bumps and clashes.

TEST NO. 3: What do you see in this image?

Figure 14 is very low quality, but do you recognize what it represents?

Do not feel bad if you do not; most people cannot figure it out... unless they live or work close to cattle. In the following image, we have outlined the main character in the photograph: it is a cow (Figure 15).

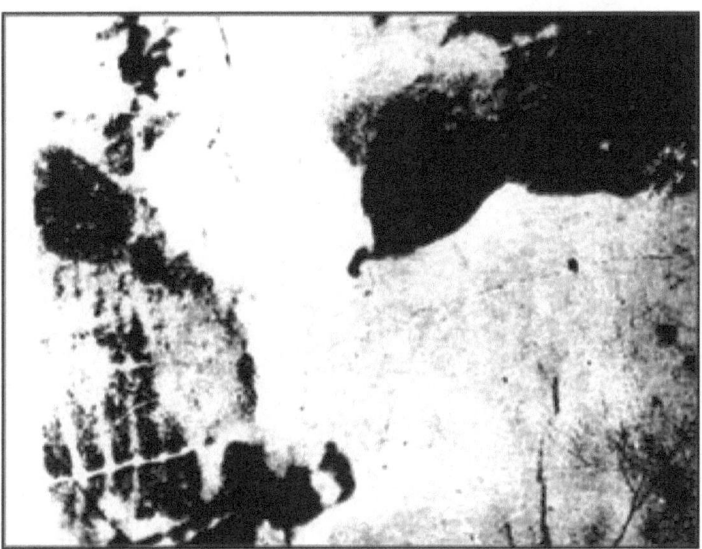

FIGURE 14   What is this? (artist unknown).

You can see it in the foreground, its head on the left side of the image and looking slightly to the right of the picture. It seems to be posing in front of a fence that we can see in the bottom left corner.

Not expecting to see a cow, it is easy to miss it. Once you know that there is a cow, it appears difficult to understand why we did not see it immediately; it seems so obvious...

The author has performed this experience many times with students and trainees. When we announce to the people doing this test that we are going to show the picture of a cow, most of them recognize it. Even the mere fact of having shown pictures of cows beforehand, or just having talked about them, makes it much easier to perceive it as such.

These extremely elaborate data processing systems allow us to optimize our perception capabilities. The amount of sensory raw information that our body sends to our brain has been estimated to be 11 million bits per second, 10 million of which are visual information (Markowsky, 2024). This vast amount of data must be structured, simplified, and screened for analysis. Our perceptual memory is in charge of the first

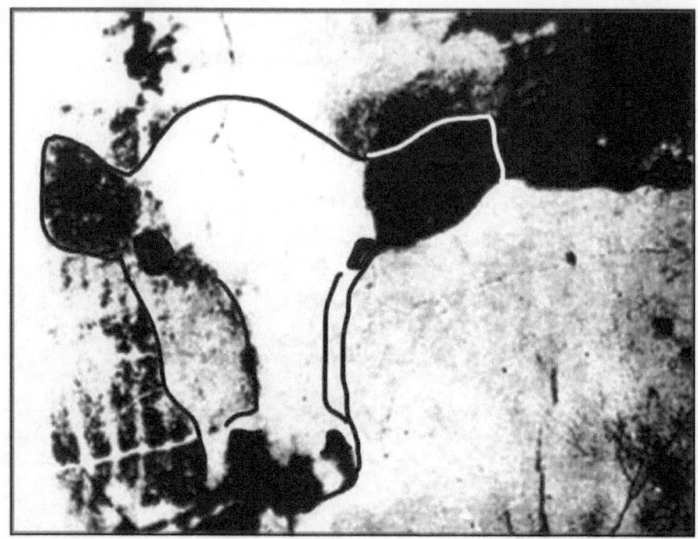

FIGURE 15   The cow (artist unknown, modified by the author).

stage of this subconscious screening process by projecting our expectations, coming from our experience, so we identify forms that have a sense in their context.

Such a sophisticated process works nicely most of the time but is, of course, fallible. **Optical illusions** are another interesting and playful way to expose this fallibility. Figure 16, known as White's illusion (Anderson, 2003), gives us the impression that the grey segments inserted in the white horizontal lines are darker than those inserted in the black lines, whereas they are identical.

FIGURE 16   White's illusion. (Adapted from Anderson (2003))

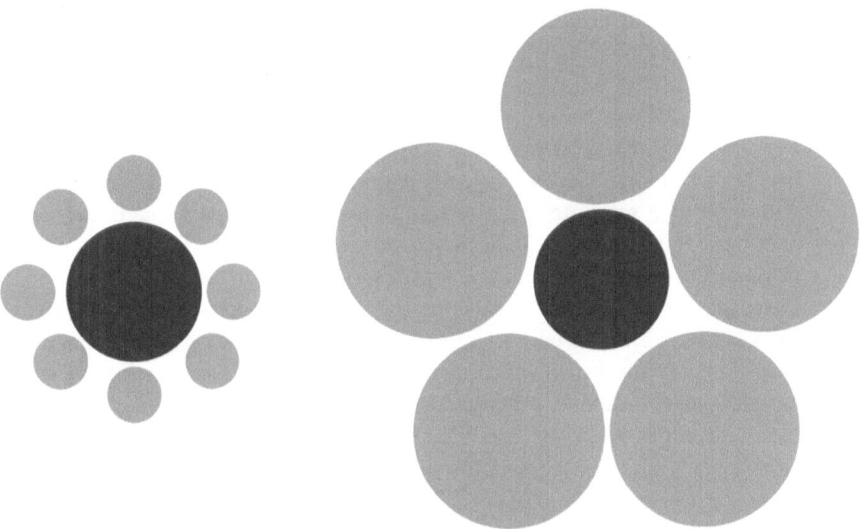

FIGURE 17   Relative size (drawing by author).

We can explain this illusion on the theories about the mechanisms of contrast perception, which assume that relative brightness is determined by the photoreceptors capturing the edge of a shape. The relative comparison of adjacent objects may also play a role, like in Figure 17: the central black circle seems bigger when placed among smaller circles (left) and smaller when among bigger ones (right), but it is the same size in both instances.

Optical illusions help us recognize the role played by the information-processing mechanisms, which operate both based on the information presented by our senses (handled by the perceptual memory) and on the basis of previously acquired information (stored in our perceptual memory as well as in other compartments of our memory that we will study later). It is by mixing all these elements that our brain constructs our perception.

We will hence use certain optical illusions to study how our brain deals with four types of visual cues that are omnipresent in our working environments and whose perception is influenced by our world's experience: **lines, colours, shadows,** and **motion**.

Lines

Lines are everywhere… especially inside our heads. Our brain visualizes them even when they are not actually drawn: we perceive as a line the divide between two surfaces presenting different optical characteristics. Unprotected ends of the rebar protruding from a neat concrete slab are particularly dangerous because our eye perceives very clearly the edge of the slab but may have trouble perceiving the rebar. So this is how we graphically represent our world, by drawing lines (and this is how we helped you recognize the cow in the test above).

We have learned to project these lines in our artificial environment, and they greatly assist us in our lives and at work to manage different kinds of risks, for instance:

- **Fencing** areas where hazards may be present. Dividing an area with a mere line on the floor – like workspaces and storage areas – or barring a path with a painted marking actually has an arresting effect. Though they are not physical obstacles, these lines become nevertheless barriers that have an influence on our movements at a sensorimotor level, as we will see.
- **Directing traffic flows**: Similar to the arresting effect described above, lines are sensorimotor cues that guide our progress when walking, driving, skiing… It is not just for decorative purposes that thousands of tonnes of paint are used for road surface marking around the world.
- **Noticing edges** and rims that may not be otherwise easily visible, like a workbench's edges or those of the stairs' steps (Figure 18).

By supplementing our stereoscopic vision, which is not very effective in some conditions (and outright non-existent at the sides of our visual field as explained above), lines also help us to better assess size and distance in our three-dimensional world

FIGURE 18 Noticeable step (picture by author).

FIGURE 19   Müller-Lyer's arrows.

by providing us with **perspective**. This impression of perspective can sometimes be misleading and cause bumps and clashes, but we can also apply it to effectively direct workers' movements (see Toolbox No. 2 – Sensorimotor Nudge).

We can use optical illusions to realize to what extent lines and their disposition have an influence on how we perceive length, size, and orientation. In Figure 19, concocted by German psychiatrist Franz Carl Müller-Lyer in the 19th century, the upper segment seems longer than the bottom one. They have the same length (you can check!).

FOCUS ON... HOW DOES THIS TRICK WORK?

You may wonder how your brain got fooled by this illusion. The trick is played by the cues displayed at the end of the segments, which subliminally simulate a three-dimensional impression.

In a very subtle way, the converging arrows placed at the ends of the upper segment give our brain the impression that this line is the *internal edge* of a volume (like, for example a room). On the contrary, the diverging arrows at the ends of the lower segment make it appear like the edge *at the forefront* of a volume. We perceive the former as being longer than the latter because our brain interprets that it is farther: if farther but appearing the same size = longer.

Figure 20 reveals the basis of Müller-Lyer's trick (the black line has the same length in both images).

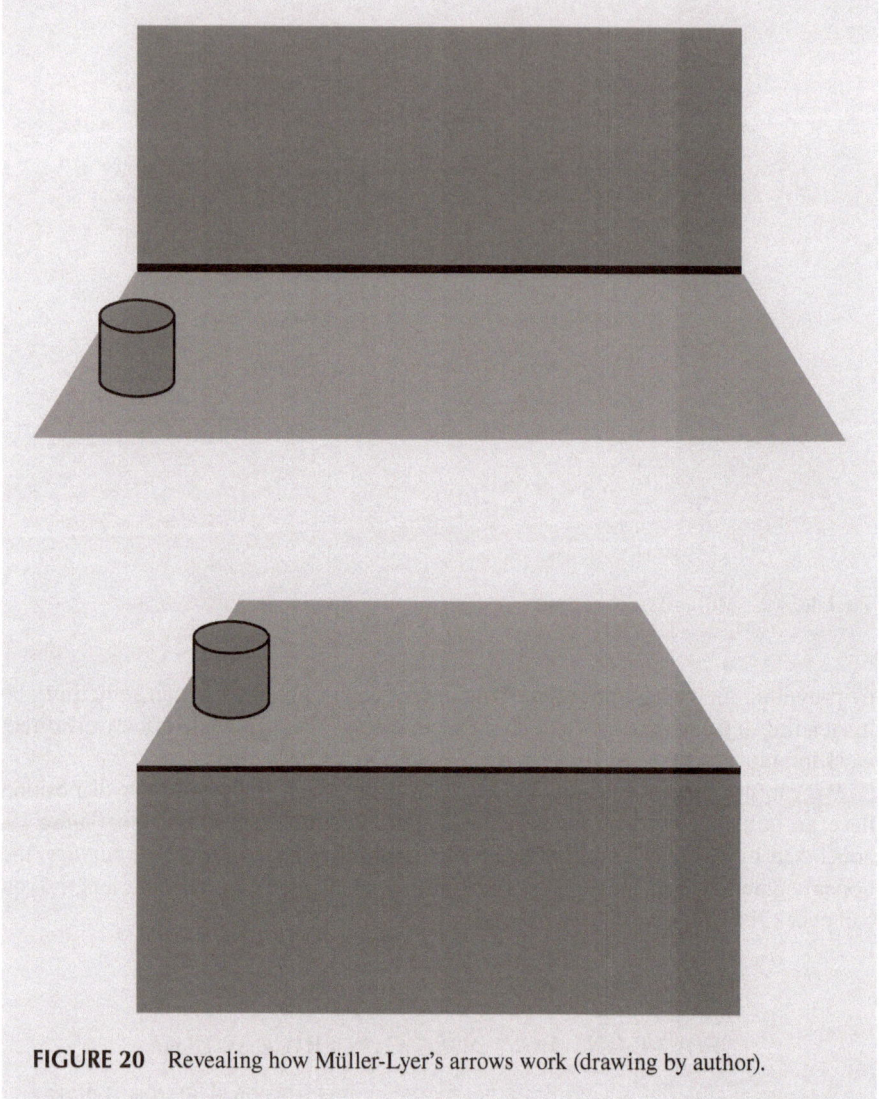

FIGURE 20   Revealing how Müller-Lyer's arrows work (drawing by author).

The same mechanism is at play in this other example, proposed by psychologist Roger Shepard in 1990. Both "tabletops" are strictly identical, but their disposition in the image suggests a 3D impression. This illusion of perspective makes our brain perceive that the left one is longer and narrower than the right one (Shepard, 1990). Again, you can easily check (Figure 21).

The simplest illustration of this phenomenon might be these bare perpendicular lines in Figure 22. They are identical in length, but the mere fact of being vertical

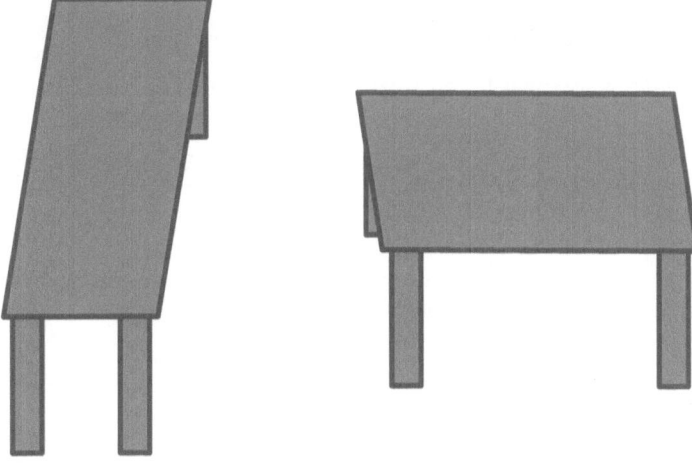

FIGURE 21 Shepard's tables.

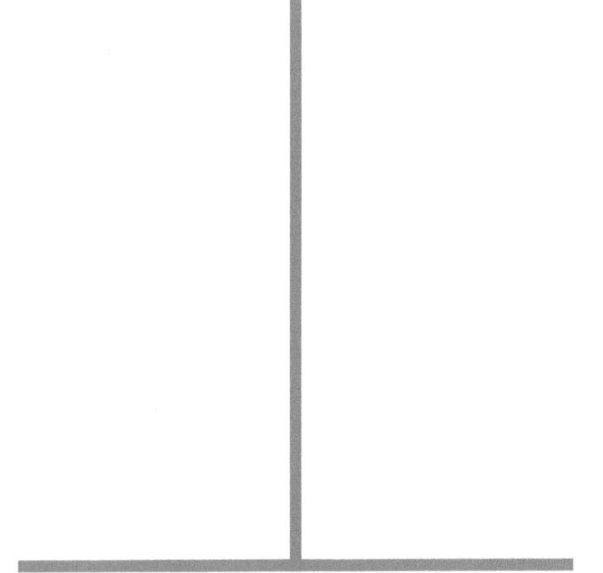

FIGURE 22 Horizontal vs vertical (drawing by author).

makes us perceive that one as longer than the horizontal one. In our natural environment, we humans are almost continually exposed to horizontal lines such as the horizon, whereas a vertical line is more often than not... a horizontal line in the ground that stretches into the distance, like a straight path ahead of us. Thus, our brain instinctively perceives vertical lines as longer than they really are.

Key Learning Points for the Safety Leader

We abundantly use lines for organizing working spaces and facilitating circulation. They constitute efficient cues that guide us even at a subconscious level.

Lack of ground marking can cause traffic accidents, and so can complex or misleading markings. Our experience of a three-dimensional world influences our brain's interpretation of lines: we instinctively integrate into them a sense of perspective that is sometimes distorted or simply illusory.

But we can use these effects for safety purposes, to subconsciously guide body movements and traffic flow, by wisely using the proper marking (cf. Toolbox No. 2 – Sensorimotor Nudge).

Colours

Even if the cones' photoreceptors in our eyes work properly (see the section on colour blindness above), our brain perceives colour in a relative, context-dependent way, which is influenced by our expectations.

Professor of vision science Edward H. Adelson published in 1995 the image presented in Figure 23, widely known as Adelson's checkerboard or checker shadow illusion (Adelson, 2005). Most, if not all, people perceive the check marked A as darker than the check marked B.

Edward H. Adelson

FIGURE 23　Checker shadow illusion (Adelson, 1995).

FIGURE 24   Adelson's illusion, unveiled (Adelson, 1995).

In fact, both squares are exactly the same shade of grey! You can fully appreciate it by masking the rest of the picture to isolate checks A and B from the surrounding context (a piece of paper with two holes will do). But just adding two vertical stripes is quite revealing, like in Figure 24.

FOCUS ON... HOW DOES THIS TRICK WORK?

Several effects combine to create this illusion. First of all, there is local contrast. Check B is surrounded by darker checks and A by lighter checks. As we saw with White's illusion above, just that may suffice to puzzle us. But Adelson's trick is more sophisticated and integrates a 3D object projecting a shadow on the checkerboard. The soft edges of this shadow are not perceived as a *line*, and in consequence, the shadowed area is not neatly visually isolated (as lines do, see above), whereas the contrast between checks is highlighted by the checks' sharp junctions.

And it is this subtle shadow that plays the trick by appealing to our expectations. In a 3D world, if check B is in the shade of the cylinder, it would receive, and thus reflect back, *less light* than other checks of the same colour but fully exposed to the light. But our eyes are getting from check B the same amount of light as they receive from check A... Since B is in the shadow and sends us as much light as A, this naturally means that B is lighter than A. This is totally coherent with our *expectations*, derived from us seeing checker-like patterns around us all the time, displaying dark and light squares aligned diagonally. B being in a diagonal of light checks, it must be brighter than A.

Key Learning Points for the Safety Leader

A lot of safety-related information is carried by **colour-coding** our visual environment. Standards and common practice in many countries have assigned meaning to colours:

- Red for signalling danger, restricted access and firefighting equipment.
- Yellow for marking potential hazards
- Blue for obligations
- Green for first aid gear and evacuation indications.

As we will see in the sections on our automatic behaviours, colour-coding safety information subliminally orients our behaviour and allows us to react faster. But as we have explained, betting too much on our ability to effectively recognize colours can be tricky. We may fail to acquire vital information in working environments that could be:

- Poorly lit (either too dark, with shaded areas or with dazzling lights)
- Dirty, smoky, or foggy
- Confusing, for instance, a messy workshop or one profusely painted and decorated
- Camouflaging, like a warehouse full of bright red barrels of oil making it difficult to quickly spot a fire extinguisher.

Shadows

Lighting not only affects colour perception but also a certain kind of cue that we constantly use in our daily lives: **shadows**.

Having evolved as a species and grown as individuals in a world where light usually comes from above, shadows constitute a very constant, hence very reliable, spatial reference. They help us locate items and determine their contact points with a surface, like in Figure 25.

Shadows also allow us to assess the volume (positive or negative) of topographic elements and more generally objects around us. A hole or a protuberance in our path can be revealed by the shadow projected, whose perception can save us from tripping and falling.

But again, the perception of shadows is influenced by our prior experience. You may have already seen Figure 26, known as the Einstein Hollow Face Illusion. Its volume is pointing away from you, but it appears to be pointing towards you. We naturally tend to see faces as convex. This bias is so strong that it trumps opposing depth cues. Seeing this artefact in 3D is even more perplexing, as the face seems to turn and follow you as you walk around it.

FIGURE 25 The shadow illusion (drawing by author).

Key Learning Points for the Safety Leader

Lighting conditions affect the way shadows of objects are projected. A poorly lit environment will naturally deprive us of this cue that, if not infallible, remains important for our spatial perception. Even when the light output is sufficient, it can mislead us if the sources of light are badly positioned, illuminating an object sideways or from below. This is a contributing factor to numerous bumps and clashes.

Movement

We tend to think that we perceive movement quite accurately. Whereas fine vision is confined to the fovea, the greater number of rods in our retina are very much able to perceive movement, even in dimly lit environments. When something moves in our field of vision, the successive activation of different photoreceptor cells makes the object *physically salient* regarding its background (we will learn more about salience in the next section).

It turns out that, even in "ideal" lighting conditions and when the moving object is fully in our field of vision, our brain is not always capable of accurately processing the information. Our evolution has made it very efficient in perceiving **natural movements,** i.e. those that we find in nature (animals, tree branches swinging in the breeze, even flowing water, although it is transparent!) and, in particular, human body movements, which we see all the time. Very subtle facial tweaks as well as the lightest nods and arm gestures are easily caught and interpreted, and even anticipated without problem.

FIGURE 26   The Einstein Hollow Face Illusion shadow illusion (artist unknown, https://www.brainfacts.org/thinking-sensing-and-behaving/vision/2013/depth-perception-and-the-hollow-face-illusion).

In contrast, we have a lot of trouble perceiving **technological movements**. For children, the inertia of the simple stick they are playing with (the simplest expression of *technology*) can surprise them and lead to an injury, for themselves or for others. With time and experience, we learn to master very sophisticated tools and machinery, but some types of motion remain difficult for our brain to apprehend. This is unfortunate, when we think about the many artefacts that move and get around in our workplaces.

The most ubiquitous of all artefacts invented by humans is perhaps the **wheel**, and all the rotating elements derived from it. The earliest archaeological evidence of wheels was dated to about 5,000 years ago, but our brain has not evolved much in the last 30,000 years. Despite our daily exposure to them and the personal experience that we gain, we are not "wired" to process this information as well as we could hope. It is hard to notice when a solid wheel is rotating (as we can be tricked by its spokes' "disappearing" when it does) and even more difficult to figure out its speed and the energy associated with it. Inexperienced workers are often surprised (and sometimes injured) by the manual grinder they are holding bouncing back.

When the worker is not holding the machine, most of the time, it is the machine's noise that warns them that an element may be rotating. But as we have seen, our hearing is not always as reliable as we can expect. Environmental noise, hearing protection devices, cochlea saturation, and subconscious override due to lack of visual confirmation can make that noise not perceived.

Some simple ideas, such as drawing spots or spokes on the side of a solid wheel, can give the operator a first visual indication that the wheel is turning. More sophisticated, but widely available, solutions include quick-acting brakes stopping the disc of angle grinders and the like.

Inertia of large masses in translation is another technological movement that our brain has not had the time to catch up with (objects weighing hundreds of pounds and hanging from slings were not something we often found in nature). Seeing such an object moving slowly, we instinctively perceive the amount of energy as being low, oftentimes low enough to try and counter the movement with our body. Translation speed can also be difficult to appreciate when an object is far away and we lack visual references, which is why planes up in the sky seem to fly very slowly.

But when translation speed becomes really problematic is when we are placed squarely **in front of the moving object**, like when we are between the train tracks or on a motorway. When in that position, we do not actually see the train or vehicle coming our way moving at all: we see it grow larger, almost imperceptibly at first, until suddenly the enlargement accelerates, leaving us very little time to react. Once again, millennia of evolution did not prepare us to accurately perceive the speed of multi-tone objects heading towards us at a hundred miles per hour, by noticing and interpreting subtle enlargement as translation.

FIGURE 27 Hurt by a wheel (picture by author).

DID YOU KNOW? STRAIGHT LINES, CURVES, MAGIC TRICKS, AND ROAD SAFETY

A magician's intuition led to a scientific experiment that proved that the eye follows more attentively the objects that move sinuously than those that move straight.

When we track with our eyes an object that follows a straight line, our eyes perform a high number of *saccades* (the quick, brief, and unconscious eye movements described above). Our eyes do not stick to the object but wander between its point of origin and the point we estimate as being its destination. However, when we observe an object zigzagging or following a curvy trajectory, our eyes remain fixed on the object. It is like, not being able to anticipate where it is going in the next second, we keep our eyes riveted on it not to lose it (Otero-Millan et al., 2011).

Sleight-of-hand experts have used this technique for a long time: they wave one hand in front of the audience so as to draw their attention to that motion, hence diverting it from the other hand that they use to do their trick.

Experienced bike riders also benefit from this empirical wisdom. They know that when drivers explore their field of vision, they are subconsciously looking for cars, not for bikes. This explains the disproportionate quantity of car–bike crashes, the driver "not having seen" the bike (we will develop this further when we will present inattentional blindness that also contributes to this phenomenon).

So, when bike riders want to make sure to catch the driver's eye, they zigzag with their bike. This may look funny to the other road users, but in any case, it is effective!

FOCUS ON... USING NATURAL MOVEMENTS TO ENHANCE DANGER PERCEPTION

Our brain's ability to efficiently perceive natural movements has not escaped advertisers, who have known for a long time that a billboard carried by a person is much more noticed than the same billboard posted on a wall.

This same principle has been used also for road safety purposes. Roadworks and dangerous zones – for instance, the area where an accident has just occurred – are much more effectively signalled by a worker with a flag or a state trooper with a flashlight than they could be by using a road sign.

Pushing this concept further, savvy safety professionals familiar with human factors sometimes put articulated dummies dressed like road workers, waving a flag thanks to a motorized device, to signal roadworks.

Key Learning Points for the Safety Leader

Our brain does not easily perceive technological movements, like rotating elements, translating large masses, and objects heading directly towards us.

In contrast, it is very good at perceiving natural movements, notably those linked to the human body. Those catch our eye and hence our attention, which is the theme of the next chapter.

TOOLBOX – Tool No. 2: Sensorimotor Nudge

We have seen that we can be easily fooled by our perception. The good news is that we can use the same tricks that sometimes mislead us to trigger safety-oriented reactions.

Let us introduce here two examples where the perspective has been manipulated to induce a speed reduction from drivers as they get close to hazardous areas. They belong to a category of methods derived from the concept of **nudge** that we will further develop later.

The first originates from Chicago, where authorities looked for a solution to the high number of road accidents occurring in a particularly wicked curve on Lake Shore Drive. They decided to paint transversal lines across the road, growing increasingly closer to one another as they approached the curve.

The effect of this disposition is that drivers approaching the curve at a constant speed get the impression that they are going faster and faster as the time to get to the following mark grows shorter, which unconsciously leads them to reduce their speed (Figure 28).

The second comes from Iceland. They drew pedestrian crossing passages that give the impression of solid 3D blocks. This also results in drivers reducing their speed when approaching the passage (Figure 29).

Driver's view   Helicopter's view

FIGURE 28 Impression of increasing speed (drawing by author).

FIGURE 29   3D pedestrian crossing (artist unknown).

ATTENTION

"Risk is in the eye of the beholder."

Diane Vaughan, The Challenger Launch Decision (1996).

It is not because a stimulus has reached our perceptual memory that it will draw our **attention**. Our brain is only capable of *consciously handling* a fraction of the 11 million bits of raw information that our senses send its way every second. A conservative estimate – considered as very optimistic by some neuroscientists in terms of order of magnitude – is that we can consciously process less than 1% of sensory data (Goldstein, 2011).

Our perceptual memory can be considered as the antechamber or entry hall of the other compartments of our memory. In this hall takes place what Treisman called the *preattentive processing* of information that, to put it simply, subconsciously sorts out the inputs on which we will focus our attention (Treisman et al., 1992).

TEST NO. 4: What do you see in this image?

You are looking at a picture of one of the façades of the National Cathedral in Washington, DC (Figure 30). If you are like most visitors walking around the building, most probably you will not notice anything weird at first sight...

As you may know, this cathedral is fairly recent; it was finished in the 1980s. When sculptor Jay Hall Carpenter was considering the design of the gargoyles, a contest was organized in the city, asking local children to

FIGURE 30 A very peculiar gargoyle (artist unknown).

propose their ideas. In the Star Wars' frenzy of the time, one of them suggested including a gargoyle representing Darth Vader. And there it is, right in the middle of the picture.

According to Triesman's **feature integration theory**, in the preattentive stage, our brain automatically gathers and registers information about basic *features*, whereas the identification and processing of particular *objects* require attention and thus happen later.

You were not expecting such a character in the façade of a cathedral. Therefore, even though your eyes captured this element, which reached your perceptual memory, it did not get through.

Before studying how our attention works, we need to understand this preliminary, subconscious, and generally unintentional process of **selectivity**. This feature is absolutely necessary for our brain to manage the bottleneck between the flow of sensory data and the capabilities of our higher cognitive functions. It is essential in our daily lives and, of course, to do our jobs effectively and safely, but it is unfortunately little known by many people.

Initially imagined as a filter (Broadbent, 1958), thankfully our brain does not pick in a random way the few inputs that it will focus its attention on. The selection is actively done on the basis of the stimuli's *salience*.

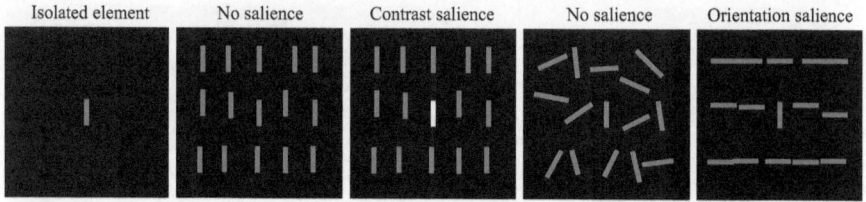

FIGURE 31   Visual salience. (Adapted from Chiaramonte (2007)).

SALIENCE

A stimulus's **salience** ("that jumps at you" in Latin), the fact that it draws our attention, is relative to the level of congruence between the stimulus and the context. Salience can be both **physical** and **cognitive**.

Physical Salience

The attributes that generate the **physical salience** of a stimulus depend on the sense concerned:

- *Sound salience* is primarily linked to its volume and frequency with regard to the background sounds or "ambient noise".
- *Visual salience* can be related to movement, relative colour or brightness (as mentioned above), as well as to shape, orientation, or contrast, like in Figure 31:

Key Learning Points for the Safety Leader

It is primarily on the basis of their physical salience that work the different types of **warning signs**, whether they are visual, audible, or even olfactory (like the mercaptans mentioned previously). We use these same visual salience attributes when we apply **visual management** techniques, which include but are not limited to those used in *lean management*.

These techniques aim to optimize the layout, arrangement and colour of signs and more generally of objects and information in the workplace. We will introduce in Toolbox No. 3 the most popular of the methods integrating this approach: the 5S (Toolbox No. 3).

Cognitive Salience

An element in our environment can also stand out from the background because of its *meaning*. **Cognitive salience** is thus determined by:

- the *level of interest* that the subject has in the particular object, and
- to what degree it *violates her expectations*,

In the absence of these two factors, occur respectively the phenomena of *inattentional blindness* and *habituation*, presented below (Figure 32).

FIGURE 32 Cognitive salience (drawing by author).

Key Learning Points for the Safety Leader

The efficacy of warnings, signage, and other visual and audible devices intended to catch the worker's attention does not depend solely on how the signal stands out physically from its environmental context. The worker must also:

- recognize its *meaning* and
- give *importance* to it, so it stands out in its cognitive context.

We will explore in detail this notion of importance when we address the concept of values.

Inattentional Blindness

Without going into the detail of the different theories on this topic yet, it is paramount to understand that in the absence of physical salience, the individuals' attention will spontaneously go to the objects that they attach importance to. Perceiving an object, an action, or a situation that is of particular importance to us will automatically draw our attention. If they lack interest to us, they become invisible to us (Mack and Rock, 1992) (Figure 33).

This concept may seem strange, yet we continuously experience this phenomenon, and there are countless instances in our daily lives that should be quite remarkable. You will certainly recognize yourself in some of the following examples:

- The day you decide to buy a particular car's make and model, you start seeing it everywhere.
- In a big family gathering, you were not paying any attention to what a person was saying until she mentions your name (also applies to professional meetings...)

FIGURE 33   Inattentional blindness (drawing by author).

- In a bar or a party, you were chatting without really listening to the music, but you instantly react when the DJ plays the biggest hit of your favourite band (if you were to say what song had been played just before, you would be unable to).

Dozens of times per day we can pass by a fire extinguisher (or a first aid kit) without noticing it, until the day we get our basic fire safety training (or first aid training). Likewise, we can walk by a hazard or dangerous situation without paying any attention if for us it is "not important". Its members spotting dangers on the go might be one of the most distinctive features of the maturity of an organization's safety culture as we will define it.

Occasionally asking someone to "pay attention" to something will, at best, have a short-term effect known by psychologists as *priming*. But such remarks will not fundamentally change their level of interest, and hence will not have a long-term effect unless they are made in a continuous and coherent manner, as we will see.

FOCUS ON... THE PRIMING EFFECT

Exposure to a stimulus can subconsciously influence our identification and response to an ulterior stimulus. This phenomenon is known as **priming**.

The first stimulus activates a region of the brain related to a part of the memory or to a representation, which remains temporarily partially activated and facilitates the perception of a second stimulus.

The priming of that second stimulus can be *direct* if it is the same stimulus. If it is not, both stimuli must be related, either *perceptually* (resemblance at the sensory level) or *conceptually* (link regarding their meaning, for example belonging to the same semantic field, by association, affective...).

For instance, it will be easier for us to notice that an operator is not wearing the mandatory gloves if we just saw the signage regarding this obligation. Or we will be more likely to notice that the floor is slippery if we just heard someone talking about the rain outside. That being said, priming is more effective if the mode of both stimuli is the same (visual, verbal, etc.).

At the very best, we can hope that generic messages delivered during the daily briefing – like "be careful" – will have a very modest effect via this priming mechanism. The more specific the message, the more effective will be the priming: for instance, "be careful today with this particular operation because of this particular danger", ideally while showing a photograph of the situation or object.

To be noted: priming can also have a *negative* effect. The fact of ignoring, consciously or unconsciously, a stimulus or message will induce a diminished receptivity to and/or a less efficient processing of associated ulterior stimuli.

Key Learning Points for the Safety Leader

The workers' training, their experiences, and their expectations will make it more or less probable for them to notice a hazard or an anomaly, or more largely any sign susceptible to bringing them safety-related information or indications.

Knowing the **meaning** of those signs and attaching **importance** to them is necessary to **notice** them.

HABITUATION

There is another very important effect of the limits of our attentional resources: we end up ignoring the stimuli that we are exposed to in a repetitive or prolonged manner. In the absence of a consequence (*reward* or *punishment* in the behaviouralist's jargon that we will present later), our brain interprets that a given stimulus is not relevant; hence, it stops paying attention to it. This phenomenon called **habituation** helps us to optimize the use of our limited attentional capacities: if it does not bite me and it is not edible, there is no need to keep spending my energy considering it (Doré and Mercier, 1992) (Figure 34).

After a while, we no longer notice the mess in the workshop, like after some minutes we no longer hear the traffic noise, and we stop noticing the reflections and the scratches in a screen. We eventually stop seeing the documents that we placed

FIGURE 34   Habituation (drawing by author).

on our desk (physical or virtual) and the mails that we intentionally left in our inbox, precisely because we wanted to keep them in view and not forget them...

Considering this, we can understand that it is when something surprises us that we notice it best. The **surprise** effect automatically allocates attention to a stimulus out of the ordinary. Surprise can be caused by the unusual intensity of the perceptive contrast (loud noise, light flash, aggressive smell) as well as by the rupture with the conceptual frame.

DID YOU KNOW? THE EXTERNAL AUDITOR

This habituation phenomenon could very well justify on its own the hassle of cross-inspections and external audits (not necessarily by a third party but at least by someone external to your site).

Beyond being the periodical stressor that makes people update their documentation, file their records and tidy things up in the workshop to "pass" the audit, having someone from outside looking at your business with "a fresh eye" is beneficial. They will be less likely to walk by situations that those who are continuously there no longer see.

This is the advantage regarding your resident auditors who, even if they surely see more issues than operational staff because they actively look for them with a trained eye and are hence less prone than them to inattentional blindness, can nevertheless fall prey to habituation.

Key Learning Points for the Safety Leader

In the absence of consequence (accident or incident, reprimand or simple remark), the repetitive and prolonged exposure to objects, situations, or signs of danger progressively reduces our ability to notice them.

Operators end up no longer seeing the indicator light that is always on in their control panel, and supervisors stop noticing the overflowing garbage containers and the bottles of chemicals forgotten in that far corner of the workshop.

However, this phenomenon is not exclusive to workers in the front line. Managers stop seeing those dashboards displaying information that they consider barely relevant and more or less outdated. Executives no longer pay attention to the "KPIs" that they receive periodically and that do not bring any lead on how to improve the situation.

CONCENTRATION

"Do not dwell in the past, do not dream of the future, concentrate the mind on the present moment".

Buddha

FIGURE 35   Concentration (drawing by author).

The selected stimuli that make it through the preattentive stage screening will enter the **focused attention stage** (Figure 35). At this stage, once attention has been allocated to an object, the features initially perceived are combined by association with the information we already possess from previous experience, resulting in the identification of the object. It is a top-down process that requires prior knowledge to identify and recognize objects. At this point starts cognitive conscious attentional processing, which can be triggered by our perception of a salient object or in an endogenous way – either spontaneously or deliberately – on the basis of internal signals from other compartments of our memory. In all cases, *emotion* plays a key mediating role that we will explore later.

After an initial preattentive stage axed on selectivity, this second stage is axed on **intensity**. At this point, we can fully consider a particular object and analyse it in depth by *inhibiting non-relevant stimuli*. The focused attention that we can hence deploy is a prerequisite for running mental simulations, evaluative thinking, systematic decision-making, rule usage, and problem-solving, as well as planning and concept acquisition, higher cognitive functions that we will address in the next chapter.

Moreover, focused attention also intervenes in perception and attention selectivity as it supports **alertness** and **vigilance**. These refer to states of mental readiness that enable us to notice and effectively process more or less predetermined stimuli.

FOCUS ON... ALERTNESS AND VIGILANCE

The concept of **vigilance** (aka *sustained concentration*) is broadly used to define our ability to maintain our attention focused over prolonged periods of time (Warm et al., 2008), deliberately mobilizing a sufficient level of attentional efficacy during the execution of a task that may be long and/or monotonous. Throughout vigilance periods, the individual actively looks for the manifestation of a particular *target stimulus*.

When in a state of **alertness**, an individual remains at the ready, in a sharper standby mode and mentally prepared to immediately react to a signal or instruction. Two kinds of alertness have been described:

1. *Tonic alertness* (aka *intrinsic alertness*) is the individual's overall level of alertness, the cognitive control of wakefulness and arousal. Quite close to the concept of vigilance, it determines our reaction time when there is no preceding warning stimulus, while providing us with the required "tone" for higher-order cognition.
2. *Phasic alertness,* on the other hand is the degree to which one can rapidly shape their response in a given instant. It implies a short attentional preparation, shorter than a second, prompted by a warning stimulus that precedes the target stimulus.

A related and interesting concept that has also been widely adopted by the human factors community and made it into the safety jargon is **situational awareness**. It integrates the two phenomena explained previously – perception and attention – and some others that we will see later, being defined as "the perception of the elements in the environment within a volume of time and space, the comprehension of their meaning, and the projection of their status in the near future" (Endsley, 1995).

Endsley's *cognitive model of situational awareness* – arguably the most influential and supported by research findings – thus proposes three levels of situational awareness formation:

1. **Level 1 – Perception:** The most basic level, it relates to detection and recognition of status, attributes, and cues.

2. **Level 2 – Comprehension**: This intermediate level of situational awareness requires the processing of the perceived elements, their interpretation
   (notably through pattern recognition), and their evaluation regarding the
   individual's goals and objectives.
3. **Level 3 – Projection**: At this level, the individual extrapolates, recognizes
   dynamics, and projects them forward in time, anticipating future possible
   scenarios, including risks and opportunities.

We can see that this model goes beyond the traditional definition of awareness as
the mere knowledge or perception of a situation or fact. To fully appreciate it, we
need to understand other higher cognitive functions that will be presented afterwards. What we will retain right away is that it also considers task and environmental factors as well as individual factors that influence situational awareness and
that, as a matter of fact, have an impact on all cognitive functions. We will deviate
a bit from traditional human factors' classifications and present them in a slightly
different way.

Physiological Factors

For our body's organs and systems to work properly, a number of physiological
parameters must be maintained in equilibrium. We call *homeostasis* the maintenance of a stable internal environment within our body, which is achieved via control mechanisms based on biochemical and endocrine feedback loops. **Homeostatic
imbalances** can affect our endocrine and nervous systems and the cognitive functions that they support.

There is a long list of potentially disabling pathologies that can lead an individual
to be declared medically unfit for a given job. We will not enter that domain and will
just focus on a series of factors that can affect us all, irrespective of our age, gender,
or health condition.

Fatigue

Physical and mental activities requiring significant efforts, whether in terms of
duration and/or intensity, result in physiological **weariness**. Factors inducing such
weariness include exposure to environmental constraints (noise, vibration, extreme
temperature…), physical constraints (exertion implying significant energy consumption or soliciting excessively our musculoskeletal system…), and psychological constraints (stress, anxiety…). Notably, an excessive *mental workload*, a consequence of
the worker using the full extent of their mental functions to perform a task, can lead
to fatigue and even exhaustion (ISO 10075-2:1996).

Fatigue impairs the workers' physical and cognitive performance in the short
term. Marathon runners midway into the race experience tunnel vision and cannot
solve even the most basic mental calculations. Workers whose jobs require a very
intense effort will face similar effects, and it is vital to take this into account when
designing tasks and operating conditions in hazardous situations.

When constraints are moderate and resting periods are sufficient in terms
of number, duration, and quality, fatigue is reversible. Our body recovers: quite

quickly after the effort, the oxygen level in our blood, muscles, and brain goes back to normal levels, the heart rate slows down, the metabolic waste is eliminated, and the amount of available glucose is restored by appealing to the glycogen reserves. Then during our sleep, our muscles are repaired and reinforced, and the nervous system is regenerated. There is no negative impact on our health, and the effects on our concentration are only temporary. On the contrary, reasonably exerting our muscles, our joints, and our brain is beneficial, helping us in developing and maintaining them.

However, when the effort is too intense and/or too frequent, it can result in negative impacts on our health over the long run. Physical and environmental constraints can lead to musculoskeletal troubles, and psychological constraints can result in disorders such as burnout and chronic fatigue syndrome.

Key Learning Points for the Safety Leader

Physiological fatigue can have negative consequences in the short-, medium-, and long-term. In the short term, fatigue has effects at two levels:

  i. *At physical level*, the accuracy and strength of the workers' moves and gestures are progressively reduced.
  ii. *At mental level*, perception, attention, and concentration are diminished.

Sparing the workers from unbearable constraints, calibrating the level of effort that they can sustainably produce, and being attentive to signs of fatigue are not just a matter of comfort for the worker and peaceful labour relations for the employer: they are a critical health and safety stake.

Energy Needs

We can easily understand that we need **energy** to perform physical efforts, but one may think that our mental activities – being of an intangible nature – do not consume much energy. The fact is that our brain, while barely representing 2% of our body weight, utilizes 20% of its total average energy consumption. This correlates with the amount of blood dedicated to its continuous irrigation (1 L out of 5 L in our body on average).

Our neurons need a constant supply of glucose. When the brain suffers a shortage of glucose (neuroglycopenia), the person becomes asthenic, and their concentration is severely impacted.

Paradoxically, just after having ingested a large amount of nutrients (typically after a heavy lunch or dinner), we can suffer from drowsiness. As a matter of fact, when digestion kicks in, a large flow of blood is directed to the stomach to provide the necessary energy and oxygen (that we will study in the section below), diverting these precious resources from our brain.

DID YOU KNOW? HYPOGLYCAEMIA AND CONCENTRATION

In the United States, 38.4 million people had diabetes in 2021, i.e. 11.6% of the population (5.6 million in the United Kingdom, i.e. 8.4% of the population). Among those, 29.7 million have been diagnosed and regulate the glucose level in their blood using drugs such as insulin (4.4 million in the United Kingdom). Exceeding the prescribed dose of these drugs may result in hypoglycaemia. A few milligrams of these substances separate the normal dose from a mortal overdose. This can be problematic if the evolution of the diabetes in the individual may require them to adjust the drug's dosage.

Other situations can have similar effects on the individual's glycaemia: skipping or delaying a meal, or having a meal without enough starches; an unplanned intense physical activity; a drug interaction...

Because the first symptoms of hypoglycaemia resemble those of high blood alcohol levels, diabetic people are advised to make their condition known to their entourage. Some decide to wear a bracelet or necklace with this information, intended to facilitate the early diagnosis of an eventual hypoglycaemia and therefore a prompt and adequate reaction from witnesses and first aiders.

Key Learning Points for the Safety Leader

The increase of diabetes prevalence in most countries around the world is a cause of concern also with regard to occupational safety, as hypoglycaemia resulting from inaccurate insulin dosage has a severe impact on cognitive functions.

While hypoglycaemia is rare among non-diabetic persons, it can occur in case of high-intensity effort during a fasting period or when a person is on a hypocaloric diet without suitable professional health-care advice.

Such questions are not easy to address with team members but can be extremely important to prevent dangerous situations.

Ventilation

The normal level of **oxygen** in our blood is usually 95% or higher. Oxygen readings below 92% are worrisome, and when these fall to 88%, immediate medical attention is necessary. Some tasks are executed in environments where fresh air may not be sufficient or in constant supply. Natural oxygen concentration in the atmosphere is almost 21% by volume. The US Occupational Safety and Health Administration regulation and several guidelines around the world set 19.5% as the lowest level acceptable for entry into confined spaces (OSHA, 1993; NIOSH, 1987).

Consciousness can be lost after just 30 seconds of oxygen deprivation. If blood circulation and oxygenation are restored within 3 minutes, recovery is likely to be complete, but in some instances, cells begin dying after just 1 minute, and after 3 minutes, lasting brain damage becomes likely.

However, without going as far as cerebral hypoxia scenarios, indoor air quality has been proven to affect our cognitive abilities. Cerebral blood flow is highly sensitive to carbon dioxide (CO_2) levels, and several studies have shown that the higher CO_2 and volatile organic compound (VOC) levels in a room, the worse test-takers do (Allen et al., 2015). Poorly ventilated workspaces, where CO_2 levels easily reach 1,200 ppm, can affect concentration and cognitive processes in general. We can find such spaces in small technical and storage rooms with no permanent occupancy and thus potentially lacking natural or mechanical ventilation, but where workers may need to spend periods of time for maintenance or inventory reasons, for instance.

During the COVID-19 pandemic, many employers intensively surveyed indoor CO_2 levels as a ventilation indicator that could be used as a proxy for exposure to airborne viruses. We realized how quickly such concentration rises in closed spaces, such as meeting rooms and individual offices. No hazardous jobs are usually performed in such administrative environments, but health issues may ensue. Coronaviruses, as well as other airborne pathogens, can cause more or less serious illnesses, and CO_2 level has been hypothesized as a migraine threshold factor (Schwarzberg, 1993). Since COVID-19 has been mentioned, many studies have been conducted on the impact of surgical and cloth face masks on blood oxygen and CO_2 levels, and they established that the feared risk of pathological gas exchange impairment related to wearing such masks is negligible in the general adult population (Shein et al., 2021; Ade et al., 2021).

We will not address industry-specific airborne pollutants that can affect workers' health and safety, as these are multiple and usually well-regulated and surveyed. Of course, intoxication by exposure to many of those substances can also affect workers' concentration, but these are normally monitored for their own sake.

Key Learning Points for the Safety Leader

Beyond the obvious and direct impact of ventilation in the management of risks related to workers' exposure to harmful airborne substances and pathogens, maintaining oxygen and CO_2 levels under control in the workspaces is important for concentration and cognitive functions in general.

Dehydration

Our body is made up of roughly 60% water, a proportion that reaches 80% in our brain and in our muscles. The first, direct effect of **dehydration** is a reduced blood volume and hence a diminished flow of oxygen that our blood carries to these, as well as the flushing out of toxins, affecting their performance. A second, indirect effect of dehydration is that despite the membranes separating them, water migrates by osmosis from our organs to the thicker, denser blood irrigating them.

Whereas dehydration of other organs can lead to health problems over the long term, impaired muscle and brain performance can quickly result in safety issues. If we focus on the brain, we can note that the neurochemical reactions and synapsis that take place within it require an extremely fine hydroelectrolytic balance to be maintained.

Severe dehydration causes muscle cramping and pain, achy joints, headaches, states of confusion, alteration of consciousness, and important neurological troubles. But even mild dehydration can affect our concentration and the rest of our cognitive functions.

Of course, tasks demanding important physical efforts and/or exposure to high temperatures will visibly result in loss of water by transpiration. Less obvious is the fact that we are losing water all the time by our respiration, in particular in dry environments (outdoors and indoors in air-conditioned working spaces). Importantly, we are less prone to notice thirst when working in cold temperatures.

Nota bene: if after reading the previous paragraph you have reached for the bottle of water that you keep at hand (a very healthy habit) you just got a glimpse of how receptive we are to subliminal influence, a phenomenon that we will explore in detail.

Key Learning Points for the Safety Leader

Being thirsty warns us that we need to drink, but thirst only appears when we are already dehydrated.

Therefore, it is not enough to supply workers with water: it is crucial to make them aware of the importance of drinking every so often even when they do not feel thirsty and to establish routines so they do it, notably via regular reminders or dedicated breaks.

This is critical for jobs and tasks involving prolonged phases of intense physical activity and/or exposing the workers to high temperatures, like construction, agriculture, welding, forging, and steelworks, as well as cooking and catering, to name a few.

As heatwave episodes become more frequent, longer, and hotter, safety risks related to dehydration will affect more workers, and in a more intense way. Adapting the tasks, the techniques, and the tools will certainly be necessary in many cases, as well as adopting specific work planning and organization.

Sleep Needs

Sleep plays a major role in recovering from fatigue, which we presented above. Not getting enough sleep has been linked with type 2 diabetes, heart disease, high blood pressure, and stroke. Furthermore, sleeping concurs in a crucial way to a healthy immune system and to the processes linked to our memory, like learning and creativity: early sleep stages benefit the consolidation of long-term memory in general, and paradoxical sleep (aka rapid eye movement or REM) participates in that of procedural memory, all of which we will study later.

Sleep is activated by a rather complex *homeostatic* and *circadian* double process.

FOCUS ON... SLEEP'S DOUBLE CYCLE

The **homeostatic process**, implying a feedback loop regulation around a suitable value of physiological parameters whose recovery depends on sleeping, can be represented as a double logarithmic curve. The need for sleep increases towards an asymptote during the time we spend awake, and then the recovery does the same during our sleep (Figure 36).

During stages 3 and 4 of our sleep, which correspond to deep sleep, our body does a "physiological cleaning" that is vital, particularly for our brain. The pituitary gland secretes during this period most of the growth hormones that will facilitate the development and maintenance of our muscles and bones, as well as the regeneration of our nervous system (Radiguet, 2018).

The **circadian process** is, for its part, responsible for the rhythmic nature of the awakening–sleeping cycle and is based on the periodical production of a series of molecules, notably *melatonin*, a hormone that prepares our body for resting and sleeping, and whose peak is reached normally between 2 am and 4 am (Figure 37).

Production of melatonin by the pineal gland is regulated by an internal "biological clock" located in the hypothalamus, by the suprachiasmatic nuclei.

Homeostatic Debt

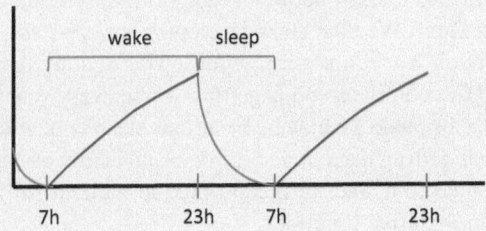

FIGURE 36   Homeostatic debt.

Circadian Rhythm

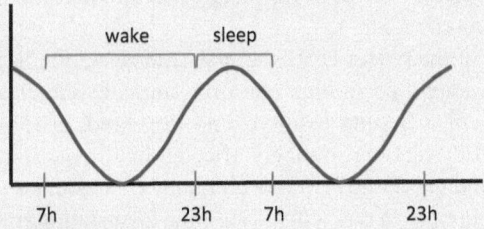

FIGURE 37   Circadian rhythm.

The electrical activity of some 20,000 neurons of which it is composed oscillates around approximately 24 hours, thanks to the cyclical expression of about 15 "clock genes".

These genes' expression is synchronized daily by the exposure to light of a particular type of photoreceptor cells in our retina: the melanopsin ganglion cells. These cells, which are wired to the suprachiasmatic nuclei by other nerves than those participating in our vision, are sensitive to blue light.

Therefore, late exposure to light, and in particular to blue light like the one emitted by our screens, will delay this clock by blocking melatonin's production. It was believed that only high-intensity light could cause this effect (in the region of a thousand lux), but it appears that light intensities as low as some dozens of lux can already trigger it. Just staring at your smartphone's screen can delay your falling asleep and then affect the quality of your sleep (INSERM, 2018).

In short, we all need our sleep to restore our physical and cognitive functions. The quantity of sleep required may not be the same for everyone, but the consensus among experts is that *a minimum of 7 hours* is advised for adults. Even if our brain does not really fully recover from a single all-nighter, if we resume our sleep hygiene after a sleepless night, its effects will fade within a few days.

Unfortunately, our lifestyles, whether chosen or imposed, make that sleeping few hours can become a habit. We then start developing a *sleep debt* – effects of getting less sleep than we should are cumulative. Almost 37% of adult Americans did not get sufficient sleep in 2022, and the trend is getting worse every year (CDC, 2024). This is a vicious condition, because we may be in serious sleep debt without always feeling tired. Similarly to our getting used to declining visual acuity as explained previously, we become accustomed to a state of mental fatigue without fully realizing to what extent it impacts our cognitive functions.

But whether we feel tired or not, our impaired ability to process new information and learn can have serious consequences for our safety at work. Our diminished focus and concentration can contribute to errors and mistakes. And drowsiness when performing hazardous activities can lead to fatal accidents. We could expect adrenaline to keep us awake in such situations, but adrenaline's boost will not help over the long run. This is typically the case for plant operators and road drivers, especially those working during the night.

Twenty per cent of the recent U.S. National Transportation Safety Board accident investigations have identified fatigue and drowsiness as either a probable cause, a contributing factor, or a finding (Marcus and Rosekind, 2015). In France, around 15% of professional drivers are routinely affected by drowsiness, and falling asleep at the wheel is behind nearly 20% of mortal highway accidents.

Whatever your opinion on daylight saving time (aka summertime), it has been statistically established that it affects road safety: a 6% increase in fatal car accidents has been reported in the United States the week after the "spring forward", an increase that reaches 8% for those living on the western edge of each time zone (Fritz et al., 2020).

When we realize that we are falling asleep when performing a dangerous activity, simply getting some rest does not help. Resting will reduce fatigue but will not remedy our drowsiness. We may turn to short-term solutions, such as consuming psychoactive substances (that we will see in the next section) or just taking a short nap, particularly indicated if caught by the after-lunch sluggishness mentioned above. Napping has been proven to offer benefits for healthy adults: it reduces fatigue and improves mood and also increases alertness and improves performance, including quicker reaction time and enhanced memory. But they are not a universal remedy, as some people just cannot sleep during the day either because of practical or physiological reasons. Napping can also have drawbacks: we can feel groggy and disoriented after waking up, and even worsen our sleeping troubles by altering our day–night sleeping cycles (Mayo Clinic, 2024).

In any case, these techniques will not erase our sleep debt and will rather be a Band-Aid. However, tackling the issue at its source is not something an employer or a manager can easily do. How many hours of sleep an employee gets is a private topic, and it is not always possible to establish a rule, however sensitive the task they perform. Few professions, like airline pilots and crews, have gone as far as to implement minimum resting hours, and it is practically impossible to actually enforce sleeping during those resting periods.

Key Learning Points for the Safety Leader

Technological devices for drowsiness detection are now available in some vehicles for surveying the driver's alertness and can be of help in other working environments.

However, preventing occupational risks related with sleep deprivation requires above all making the workers aware of the importance of sleep for their health and safety. This includes both operators and managers: the former must review their lifestyles and the latter their planning of schedules, shifts' organization and vigilance regarding persistent signs of fatigue in the team's members.

If a potential issue is detected, whether by a manager or a colleague, it must be addressed in a benevolent manner. Doing so will be all the easier if we have managed to develop a climate of trust, a topic that we will examine later.

Psychoactive Substances

Number of **psychoactive substances** have an immediate impact on our cognitive functions: some of them have mild effects and others have much more serious consequences. Consuming them regularly can also cause long-term effects and conditions. Depending on the country and the state, some of them are legal and others are not. In any case, the prevalence of their consumption depends on the local culture as well as on the age and social influences. Below we present a shortlist of the most common ones.

Legal Drugs

There are many legal drugs that alter our perception, attention, and concentration. Medicine doctors must warn their patients if the drugs they prescribe can have such effects, but some of these drugs may be bought over the counter in many jurisdictions. For example this is the case of antihistamines used to fight the symptoms of allergies and some cough syrups.

But the two types of legal drugs that have the most severe effects on our cognitive processes are:

- **Anxiolytics** (aka anti-anxiety drugs) and **antidepressants** such as *benzodiazepine*, used by 30.6 million adults in 2015 in the United States (12.6% of the adult population). This figure includes 25.3 million (10.4%) as-prescribed and 5.3 million (2.2%) with misuse (Maust et al., 2019).
- **Painkillers**, and in particular **opioids** (whether natural, also called *opiates*, or synthetic). Opioids are used by 22.1% of adults with chronic pain in the United States in the last 3 months in 2019. Considering that an estimated 20.4% of US adults suffer from chronic pain, almost 5% of the adult population is regularly prescribed opioids in the United States (NHSR, 2021).

Other Legal Substances

Alcohol   According to the 2023 National Survey on Drug Use and Health, 132.9 million adults ages 18 and older (51% in this age group) reported drinking alcohol at least once per month in the United States, and 16.3 million (6.3%) were heavy drinkers (NIAAA, 2024). Being an anaesthetic substance, alcohol's immediate narcotic effects slow down communication between neurons, notably in the hippocampus (hindering the passage of information from working memory to long-term memory), the frontal cortex (affecting rational thinking), and the cerebellum (impacting our motor skills). This results in uninhibited and reckless behaviours combined with a lack of coordination, a too-often fatal mix.

Alcohol acts as a sedative, and when large amounts of alcohol are ingested before bedtime, sleep onset latency is decreased and sleep architecture is altered. We may think alcohol helps us rest as sleep comes easier, but it is of poorer quality (Colrain et al., 2014). In the long run, regular alcohol consumption modifies the anatomy of the brain as it loses tissue (after 10–15 years, a heavy drinker's brain can lose 15% of its volume) and alters its functions.

Nicotine   Once inhaled via tobacco smoke or vapours from electronic cigarettes, nicotine reaches the brain in just a few seconds, triggering anxiolytic and stimulating effects.

It also induces vasoconstriction that limits the flow of blood, notably to the brain, and can result in light-headedness and weakness. Reduced flow of blood to fingers can contribute to their numbing, especially in workers exposed to cold temperatures. This can lead to diminished dexterity and also prevent the individual from feeling the pain of a cut or scratch, which could get infected if left untreated.

Being highly addictive, the effects of withdrawal – if less severe than those of alcohol and other drugs – can be problematic. Common withdrawal symptoms include urges, cravings, anxiety, restlessness, and difficulty concentrating, as well as sleeping troubles (and we have seen how important sleeping is) and feeling irritated, cranky, or upset, which can lead to communication and collaboration issues with management and other team members.

Caffeine   Present, of course, in coffee but also in tea and guarana – and in massive doses in energy drinks – caffeine also reaches the brain shortly after ingestion. It has stimulating effects and facilitates perception, attention, and in general cognitive functions as it reduces reaction time. Recent studies have shown that it accelerates the hippocampus's response to tasks involving memory, hence helping learning processes notably when consumed regularly (Nehliq, 2012).

However, it has also been noted in regular caffeine consumers a relative decrease in neural activity when they have not gotten their usual dose. Conversely, high doses (and even moderate doses in people particularly sensitive to the molecule) can result in excitation and even anxiety. Furthermore, caffeine intake delays sleep onset and thus reduces sleeping time and also its quality.

Illegal Substances

Cannabis   Tetrahydrocannabinol (THC), the psychoactive substance contained in cannabis, augments the brain's dopamine levels, thus inducing a sense of pleasure, but also alters perception, sensorimotor skills, memory, and judgement.

Although legal in some countries and states, cannabis consumption is banned in most of the world. In any case, a 2023 Gallup poll reported that 16% of adults in United States consume marijuana regularly (Gallup, 2024). Cannabis is the most commonly used illicit drug in the United Kingdom, with 7.4% of adults 16–59 years old declaring having consumed it in the past year (ONS, 2022). In France, a country where cannabis is also illegal, this proportion attains 31% among construction temporary workers, a profession where the reduction of alcohol-related problems has been more than offset by the increase in troubles related to marijuana (FFB, 2024).

Cocaine   The second most used illicit drug in the United States behind marijuana, 5.3 million Americans used cocaine in the past year in 2022 (SAMHSA, 2023). Cocaine blocks the recapture of certain neurotransmitters linked to the feeling of pleasure, such as dopamine as well as serotonin (which causes a self-confidence effect) and noradrenalin (which has euphoric effects), leading to their accumulation in the brain.

This causes immediate sensory stimulation that comes with disinhibition and a feeling of invincibility that, like alcohol, often leads to inconsiderate risk-taking.

Extremely addictive, cocaine use triggers a vicious behavioural feedback loop: the downhill phase causes feelings of exhaustion, anxiety, and sadness, which the users will try to escape by consuming a new dose, notably to be able to cope with their employment's constraints.

Opioids   Already mentioned above as ingredients in painkillers for their analgesic and sedative properties, opioids are also very powerful euphoriant substances, and many of them are among the most addictive drugs, causing extreme dependence. Globally, opioids are the second most used illegal drug, with some 60 million people in the world using them for non-medical reasons at least once in 2021 (UNODC, 2023).

The United States are currently going through an "opioid epidemic" of devastating human and economic consequences. Use of illegally made fentanyl and fentanyl analogues rose sharply as from 2013, then exploded in 2019. Of the nearly 108,000 deaths from drug overdose in the country in 2022, some 82,000 (more than three quarters) involved opioids, mostly fentanyl and its analogues (CDC, 2024). One can only guess the impact of this situation on occupational accidents.

Key Learning Points for the Safety Leader

In most countries, medical confidentiality forbids employers from inquiring about the workers' conditions and the drugs that may be prescribed to them.

Therefore, it is by raising awareness on the potential effects of prescription drugs on work safety (combined with the development of psychologically safe workplaces, a topic that we will address later) that management can bring workers to declare the kind of medical treatment that they are following.

Handling the matter of alcohol and illegal substance usage, whether due to addiction or occasional use, is even more complicated, all the more so that such usage may be linked to attitudes that are rarely compatible with precaution, safety, and care. In some jurisdictions and/or for some specific jobs, in case of suspicion of influence, employers can test the workers and/or appeal to law enforcement (or may even be required to). However, in many cases, employers can only forbid the use of these substances in the workplace and rely on management to have these bans respected.

TASK AND ENVIRONMENTAL FACTORS

As a matter of fact, it is not possible to draw a line between **task- and environmental-related factors** and what we previously called physiological factors. We have already spoken about fatigue, which can be "imported" by the individual from their private life but can also be produced by the task. The ventilation of the workspace is obviously related to the job and the place where it is performed. Dehydration is also very much connected to the task and its environment. And even sleep troubles and disorders are often derived from, or aggravated by, work-related worries, stress, and anxiety.

Task- and environmental-related factors will also have a crucial role in other cognitive processes that we will address down the road. The ones that we will now see are rather linked to the nature of the information that the workers will need to process, how it is presented to them, and how their attention will handle them – thus without yet addressing the topic of logic and other psychological matters that we will study in the chapter about our mind.

Mental Workload

Mental workload is a concept that is at the heart of human factors and ergonomics studies. The rhythm of technology adoption in many working environments over the 20th century, while reducing physical demands, outpaced in many cases our ability to cope with ever-increasing cognitive demands (Young et al., 2015). When task complexity exceeds the worker's cognitive capacity to meet the task's demands, we can be certain that "errors" will ensue.

In the following sections, we will review in detail the impacts of complexity on higher cognitive processes and our ability to handle it. At this stage, we can already observe that excessive cognitive burden impairs our capacity to effectively focus our attention and concentrate. However, *work underload* can lead to monotony, as explained below.

Key Learning Points for the Safety Leader

Properly assessing the **mental workload** imposed upon an operator by the task and the working environment is a prerequisite for job design and the evaluation of the related health and safety risks at many different levels, and its relevance is not limited to make sure that the task's demands do not exceed the workers' attentional span and concentration capacities.

However, oftentimes such evaluations are neglected, resulting in gaps already at this stage of the worker's cognitive processes. Whether they are ill-informed or just optimistic, engineers and managers tend to put too much stock on the operator's ability to cope with inordinately large amounts of cognitive demands.

Monotony

Some may say that, fortunately, not all jobs constantly require the full concentration of the operator and then some more. And some may even consider that the occasional moment of "dullness" is something that some workers could appreciate as relaxing. That might be just fine, as long as there are no hazards involved.

It is extremely difficult to keep one's attention focused on a **monotonous task**. Monotony is related to subjective factors such as job interest and satisfaction, as well as to objective working conditions. It has been extensively studied in driving and has been proven to impact concentration in short-cycle repetitive work (Melamed et al., 1995). Prolonged monotony slows down circulatory function, the respiratory system, and brain activity, strongly intensifying drowsiness (Thiffault and Bergeron, 2003). At the brain level, this is related to *habituation*, which we have already discussed, and results in a progressive reduction in alertness, vigilance, and situational awareness.

Our attentional resources are spontaneously directed towards more salient objects or towards internal mental processes (e.g. daydreaming), and if we are performing a task, its execution becomes *automatic* (a topic that we will revert to later).

DID YOU KNOW? HIGHWAY HYPNOSIS

Taking a break when performing monotonous tasks allows for physical and mental rest, but above all it allows the workers to refocus their attention. The advice to make a stop every couple of hours when driving is a good example. It is not just because we need to stretch our legs and relieve our bladders that taking these breaks is a good idea: they permit us to clear our minds and prevent what has been called **highway hypnosis**, a term coined by G.W. Williams in 1963 (Underwood, 2005).

Indeed, to remain seated, still, staring at the road ahead without much to do but holding the wheel, can bring us to a state of stupor in which we lose contact with what is going on around us. In some instances, this trance state can be so deep that visual and auditory distortions occur. This is particularly frequent when driving out of town, as there are fewer attentional solicitations, and is, of course, more difficult to manage at night, sunset, and after lunch (the periods when we are naturally affected by drowsiness).

Key Learning Points for the Safety Leader

Certain activities are intrinsically monotonous, like long-distance driving, and naturally result in the workers' inability to remain focused on the task.

Professional drivers are advised (and even mandated by law in some countries) to take periodical breaks. In industrial settings, we can imagine alternating different tasks and even workers' rotation between different workstations, which requires developing the workers' polyvalence.

Monotonous tasks can also lead workers to actively look for a way to keep their mind busy, and hence for *distractions*, which, as we will see in the section below, bring their own lot of problems when it comes to safety.

Distractions

The capacity to inhibit non-relevant information that characterizes the concentration of our attentional resources (supported by the *selectivity* and *intensity* functions already discussed) is coupled with a third essential executive function: our attention's **flexibility**, which allows us to switch between targets and distractors.

Our ability to willingly focus our attention is sometimes referred to as **top-down attention**, as opposed to **bottom-up attention**, which is automatically "caught" by salient stimuli, as we saw earlier (Burchman and Miller, 2007). Both types of attention are necessary for our safety, as we must simultaneously:

- Be able to focus our attention and concentrate when performing a delicate task, and

- Be able to pick up a signal coming from outside the focus of our attention, which may warn us of an unexpected potential danger.

However necessary, these two types of attention can also be sources of trouble. Both can act as **distractors** and prevent the other to effectively direct our attention for our safety's sake:

- Be "too much concentrated" on a task to the point of not noticing a sign of danger, or inversely
- Be perturbed by a stimulus when we must focus all our attention on the task (in particular by stimuli to which we *react by reflex*, such as a ringing phone, a concept that we will explore later).

Key Learning Points for the Safety Leader

Potentially dangerous tasks and working environments must be designed so as to prevent the operator's attention from being diverted during execution while ensuring its availability to detect warning signs and other kinds of stimuli bringing essential safety information.

Examples of practices concurring with these objectives are:

- Drills and exercises to ensure that sirens and other evacuation signals are effectively perceived, despite the ambient noise and eventual hearing protection devices.
- Obligation to set the smartphones in flight mode while driving or performing safety-critical tasks.
- Banning personal smartphones from workshops and construction sites.

However, like for the management of monotonous tasks, there is no catalogue of off-the-shelf solutions that could apply to multiple kinds of jobs and job settings. Possible distractions must then be investigated on a case-by-case basis during the **job safety analysis**, which must include a detailed review of the tasks' attentional demands, including but not limited to distractions, to ascertain that they can be matched by the worker's attentional resources.

Divided Attention

Modern technology in general, and information and communication technologies in particular, make it so that we are overloaded and overwhelmed by attentional demands. The number of stimuli we must tend to has been multiplied in just a few decades, whereas our brain has not changed much in the last 30,000 years as we mentioned previously. Our prefrontal cortex and our attentional span evolved in a natural and social context where doing one thing at a time was enough: walking, hunting, eating... Our cognitive processes occur in the frame of this "wiring", and we have adapted to manipulate information in a serial manner.

Many different experiments have been conducted to measure our ability to perform more than one task at the same time, and a certain amount of controversy still exists (Bouchard et al., 2025).

TEST NO. 5: Multi-tasking

Are you among those who think that they can do two things at the same time? Try to read the text below, from top to bottom (Figure 38).

Quite quickly we remark that there are two messages here – one sentence written in darker font and another in lighter font – and that our brain is not capable of processing them both simultaneously.

Keep that in mind next time that you are in a meeting, presenting your project to colleagues who keep looking at their screens...

> **You have four times**
> it is very hard
> **more chances to have**
> to mentally handle
> **a car accident**
> two things
> **when you are having**
> simultaneously
> **a phone conversation.**

FIGURE 38   Multi-tasking test.

For our practical purposes, and in real-life situations, we must stick to the abundantly proven principle that our brain cannot effectively manage two different flows of information simultaneously (Pashler, 1994).

However, laypeople persist in believing that they can execute two or more tasks in parallel, an "ability" that has been called multi-tasking (Sanbonmatsu et al., 2013). Two other (and real) abilities concur to give us this impression:

- With practice, we learn to *switch* rather quickly from one task to another. For instance, when driving we can take a look at our GPS and promptly revert back to the road, and think that we have effectively got all the information necessary for us to drive safely. We just do not realize how many things we have missed while our eyes (and our attention) were away from the road.
- We can perform one task that requires our attention while simultaneously executing another task in an *automatic mode*. We are under the illusion that we have conscious control over both tasks, but this is not the case. When we

are behind the wheel, we can have an animated conversation with our passengers or on the phone and "drive" at the same time. Indeed, we can keep the vehicle's trajectory and a distance from the car ahead, hit the brakes if that car's brake lights turn on, and even take the right turns if we know the itinerary by heart. But all these are fully automatic behaviours, which we will analyse later, over which we have little conscious control.

There is no danger for an operator to run two different machines or to survey a control panel with many instruments on it, as long as the person:

- Can tend to all these in a sequential mode,
- Does not need to focus their attention on more than one thing at a time, and
- Will be warned by a salient enough signal if their attention must be redirected at any time (as explained in the section on *bottom-up attention* above).

But to think that someone can make a complex setting machine while "keeping an eye" on what is going on next door is just a dangerous utopia.

DID YOU KNOW? MULTI-TASKING CATASTROPHES

On July 24, 2013, a high-speed train derailed just a few miles from its destination in Santiago de Compostela (Spain), killing 79 passengers. The train was circulating at 179 km/h (111 mi/h) when it entered a sharp turn where speed was limited to 80 km/h (50 mi/h).

The train driver was very experienced and knew well the itinerary. Just when he was to start reducing speed, he got a call from a member of the train crew and got into a discussion over which platform they needed to go to.

Key Learning Points for the Safety Leader

The popular belief on our ability to effectively split our attention between different tasks is behind many accidents, on the road as well as in the workplace. When one of these tasks implies important risks, a lack of attention at the wrong moment can lead to fatal consequences.

Beyond providing awareness of this natural limit of our brain, employers must organize the tasks and engineer the methods and means so as to avoid workers being tempted to optimize their time by doing several things simultaneously.

Because more often than not, this tendency to engage in different tasks at the same time is born from the desire to be as "efficient" as possible. Therefore, we will need the three pillars of safety (organization, technology, and human in the form of awareness) to overcome our natural tendencies and make operators and management willing to take their time.

MEMORY

"Memory is always at the heart's command."

Antoine de Rivarol (1753–1801)

Our memory is organized around five different functional systems, which are inter-connected. We have already studied our perceptual memory. We will now review the four others: working, procedural, episodic, and semantic memory.

Working Memory

The attentional span's limits that we have described above stem from the inextensible capabilities of our **working memory**. Focusing our attention on a selection of stimuli amounts to interiorizing them in this second compartment of our memory, which allows us to temporarily store and mentally manipulate them. This is the basis of cognitive processes such as understanding, reasoning, and learning.

Derived from the *short-term memory* concept (Broadbent, 1958), working memory abandons the notion of unitary storage and replaces it with a multi-component system. But above all, it integrates this system within the whole cognitive system, with a central executive control of our attention (Atkinson and Shiffrin, 1968; Baddeley and Hitch, 1974; Baddeley, 2000).

Beyond the nuances that each theory brings, our working memory is characterized by three inherent limitations: physiological, cognitive, and situational.

Physiological Limitations

By the very nature of the neural circuitry allotted to it, our working memory has two intrinsic boundaries:

- **Limited number of items**: our memory span is defined as the longest list of items (numbers, words, letters, etc.) that one can repeat back in correct order immediately after presentation on 50% of all trials. On average, we can retain between five and nine, very rarely more. The median being seven, this figure has become popular as "Miller's magic number" after George Miller's famous paper in the 1950s (Miller, 1956).
- **Limited duration**: these items cannot remain in our working memory for longer than some 20 seconds, unless we "refresh" them either physically (by repeating the exposure to the stimuli) or mentally (by rememorating them).

Key Learning Points for the Safety Leader

The way instructions are given to the workers must take into account the natural boundaries of their ability to retain and mentally manipulate information. This ability will vary between individuals and will also fluctuate for the same person depending on factors affecting their *concentration*, as presented previously.

When necessary, a strategy to pass information into their long-term memory must be formulated, either by the individual or by the organization, as we will see later.

Cognitive Limitations

Like perceptual memory, our working memory has a lot of trouble integrating unknown or incongruous information:

- Our *prior knowledge*, all that we have stored in our long-term memory, is used as a sorting key to screen the information that we perceive and draw our attention to what is familiar to us. It can also deform this information to adjust it to our world experience and learnings (in Goethe's words: *We only see what we know*).
- Our *expectations* and *anticipations* do the same (or as Madonna sang it: *You only see what your eyes want to see*)

TEST NO. 6: Perception of changes

Look attentively at this image in Figure 39, then turn the page and look at the image in Figure 40.

Without looking back at the previous image, can you identify the four differences? They are identified in the image in Figure 41.

FIGURE 39   Seven differences, 1 on 3 (picture by author).

FIGURE 40 Seven differences, 2 on 3 (drawing by author).

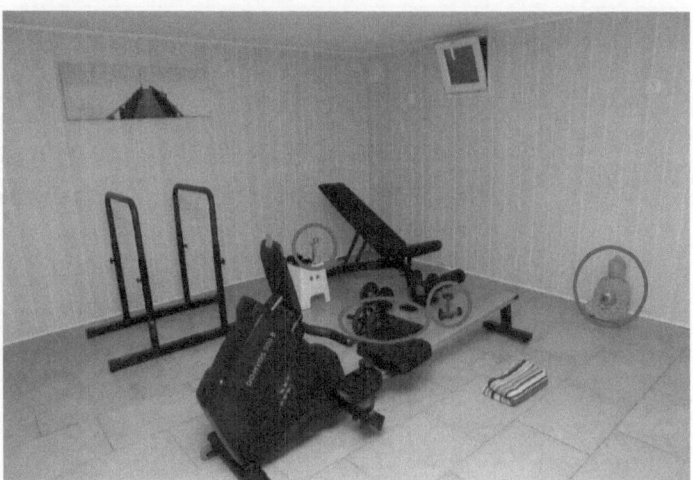

FIGURE 41 Seven differences, 3 on 3 (picture by author).

This kind of game is particularly difficult for us. They allow us to illustrate and integrate several of the points raised so far:

- Should the image change before our eyes, our visual memory would detect the *physically salient* modifications. But the break in the visual continuum (turning the page) prevents this from happening.
- So, we call on our working memory. However, since most of these objects do not have a priori any particular meaning to us, it is unlikely

that you internalized them at that level – unless you used a mnemonic technique like the ones we will present later. If, on the first image, you had seen your national flag and, in the second, that had been substituted by a foreign one, you would have most probably spotted the switch. Or if we had swapped your car brand's logo for another one: objects that have meaning to us "jump at us".

- Not only these objects are meaningless, but they are also displayed in total disorder. There is no pattern whose disruption could help us spot a change. If the objects had been classified by colour or size, it would have been easier to notice that one was no longer in the right place.

This is a "paper version" of what is known as the *flicker test*, used to demonstrate our **change blindness**. We are surprised by this limitation because – as we saw in the section on perception – we usually perceive movement in our visual field rather well. But if there is an interruption in the visual continuity, like looking away for a second or even just blinking, such ability does not help.

Key Learning Points for the Safety Leader

Personnel induction, awareness, and training schemes are not just instruments for teaching the workers the basic competencies to perform their jobs: they must also provide them with the information and skills to recognize the cues and signals indicating dangers and anomalies that they might be confronted with one day, even (or especially) if such dangers and anomalies are only susceptible to happening on an exceptional basis.

Drills and practical exercises are the most efficient way to deliver this information, as we will see. But whatever the form that may take the vectors of this information, they must be continuously refreshed.

Situational Limitations

Finally, the environmental context will influence our working memory's efficacy in handling the information. The three following factors will have a negative impact on this processing:

- **Interferences**: *attentional flexibility* and the ensuing *distractions* (as we saw above) are linked to our working memory's *receptivity* towards different information flows. These flows can interfere with each other and in particular with the one we intend to focus on.
- **Interruptions**: a disruption while performing a task is equivalent to resetting our working memory. The cognitive effort required to mentally get back on track is tiring, and if interruptions are frequent, they can result in diminishing motivation to concentrate, frustration, and even stress (next point).

- **Stress**: when the workers perceive a discrepancy between the task's requirements and the means they possess to fulfil it (time, tools, competence,
  skills…), stress kicks in, triggering a series of physiological, neurological
  and psychological processes that will negatively affect their working memory and that we will come back to.

These same situational limitations can also impact tasks controlled by our *procedural memory*, which we will address in the following section.

Key Learning Points for the Safety Leader

Organizing work by using checklists and other procedures and instructions,
broken down into steps, helps to reduce stress levels and recover from interference and interruptions.

Although this form of "information storage externalization" may inhibit
long-term memorization by the individual (as we shall see later), it may prove
necessary in the case of complex operations and/or operations involving major
risks, as in the aeronautical and nuclear sectors.

Since it is not always possible to impose working in pairs or constant supervision – which may nevertheless be required in certain cases where the risks
are major – this documentary approach commands extreme rigour on the part
of the worker. This is why it can only work in organizations with a mature
safety culture (more on this later).

PROCEDURAL MEMORY

Our **procedural memory** – hosted in different brain regions such as the cerebellum,
the basal ganglia, and the striatum – is a type of implicit (non-declarative) long-term
memory: it does not require conscious thought for either encoding or retrieval, and
it is stored lastingly.

The gestures, actions, and behaviours we repeat in a given context end up being
integrated into this memory, which enables them to be reproduced *subconsciously*:
they become **automatic**. The underlying mental processes are no longer managed by
the working memory we have just seen. This substantially increases the speed and
precision of execution while at the same time freeing up the working memory, for
example to plan ahead for other tasks that may need to be done in the future.

DID YOU KNOW? NEUROPLASTICITY

Using functional MRI techniques, neuroscientists studied the brains of individuals during intensive learning of various disciplines, from juggling to violin.
They were able to observe an increase in the size of the brain areas involved in
that practice, notably in the regions mentioned above.

This **neuroplasticity** of the brain, already mentioned with regard to the perception by blind people, allows it to increase the number of synapses and to increase the diameter of the fibres, thus accelerating the propagation of information and facilitating the development of automatisms which improve performance.

With practice, neural pathways gradually become roads, then motorways (Fino, 2007).

We have all seen striking examples of the effectiveness of this "automation by repetition", which can even astonish us, like top sportspeople and top musicians. But our everyday lives are also full of these automatisms, which more often than not go completely unnoticed. Through experience and practice, over the course of our lives we build up an immense repertoire of sequences of gestures and actions – as well as verbal and non-verbal expressions – that we reproduce without thinking, either individually or in combination, when a trigger stimulus is identified by our brain.

Unfortunately, this gain in speed and precision comes at a cost: the loss of conscious supervision over the task. This is a counterintuitive phenomenon that may surprise us, given our impression that we are constantly "in control of our actions", but it is one that needs to be integrated into the design of processes, resources, and work environments.

In a workshop, when a worker needs her tool placed on her workbench, she looks for it with her eyes (or not...), stretches her arm in its direction, and then flexes her fingers' muscles to grab it without any thought or conscious calculation. If one day someone – including herself – were to place a sharp object on that same workbench, in the trajectory that her hand usually follows, she would have every chance of injuring herself.

Three kinds of automatic behaviours deserve particular attention: our sensorimotor intelligence, our conditioned reflexes, and our procedural automatisms.

Sensorimotor Intelligence

The spatiotemporal coordination of gestures and motion is essentially managed by the cerebellum, which integrates signals from the spinal cord with those from several areas of the cortex and synchronizes the different motor areas during our movements. With experience, we walk and, more generally, we use our limbs to carry out everyday gestures without any need to think.

DID YOU KNOW? MIRROR NEURONS

In the 1990s, a team from the neuroscience department at the University of Parma discovered neurons in various regions of the macaque's brain that activated both when it performed certain actions and when it observed another individual performing them. Neurons acting in this way have since been observed in certain birds (neurons that activate both when the bird sings and when it hears a fellow bird singing), and in 2010 their presence in our brains was confirmed (Keysers and Gazzola, 2010).

These **mirror neurons** subconsciously push us to reproduce certain gestures, such as yawning or the posture of the person we are talking to, and they make us follow the trajectory of the person or vehicle ahead of us in a queue.

Mirror neurons also help us learn to reproduce movements by watching others do them, even if perfecting the ability to perform them fluidly requires practice.

They are also thought to be related to empathy, making it easier to understand the emotions of others, a phenomenon we will discuss later.

One of the remarkable results of this sensorimotor learning is the very strong link that forms between gaze and gesture. When we are driving a vehicle such as a car or a bicycle, the mere fact of staring at an obstacle in our path (a pothole, for example) causes us to direct our trajectory towards it. In the same driving situation, diverting our gaze to one side of the road to admire the landscape or to consider the message on a billboard causes us to drift towards the side we are looking at.

This eye–gesture connection is why, whether you are at the wheel, riding a bike, or carrying out any dangerous task, it is vital to keep your eyes focused on the target of your gesture. When we learn to ride a motorbike, we are taught not to stare at the pothole but at the path we will have to follow to avoid it, not to look at the outside of a curve but to the point we want to reach at its exit.

The same applies to situations where the operator is using a tool (like drilling, welding, hammering…). It is the fact of locking the target with one's eyes that ensures that the dozens of muscles and hundreds of nerves involved are activated and coordinated without conscious thought. This is also what we do when we aim at a target, for example with a bow and arrow: we keep our gaze locked on the bullseye until the projectile has gone off. Not because the target might suddenly move, but because that's how we align the activation of the muscles involved in that shot. If our eyes deviate from the target, our gesture will most likely deviate too.

This is another vital reason to avoid *distractions*, a theme that we have already addressed, which may make us look away from our target when performing a delicate gesture.

Another phenomenon that is particularly important for our safety also relates to the gaze–gesture link: when we move, we have automated the management of this link so that our eyes will, without any conscious calculation on our part, look to the point that provides us with the most useful information for adjusting our trajectory.

In fact, as we move forward, *we automatically look at the point we are going to reach in about 3 seconds.* When we are driving a car at 130 km/h (80 mi/h) on the motorway, we set our gaze about 100 m (110 yards) from the car's hood, whereas when we're walking our eyes naturally focus on the path some 3 m (3.3 yards) from our feet.

This helps us to avoid obstacles but can cause accidents if we are, again, *distracted*. If a stimulus attracts our gaze as we move forward and we deviate our stare from our path, in many instances we do not stop on our tracks but we continue to progress. Then, when we regain visual contact with our path, we will automatically direct our *central vision* at the point we will attain in 3 seconds, and not the last point we saw on our path before looking away.

This means that if, while walking at 4 km/h (2.5 mi/h), we look away from our path for 5 seconds, we will have missed more than five and a half metres of it (5.5 yards). When driving at 130 km/h (80 mi/h), that distance will be more than 180 m (almost 200 yards).

Even if it is within our visual field, that part of our path will hence only be covered by our *peripheral vision,* which, as we saw, has a rather poor resolution. Not having set our central vision on it means that it is unlikely that we will spot a cable, an oil puddle, a hole, a step... or anything else unless it is bright, high contrast, or moving.

Key Learning Points for the Safety Leader

Our ability to manage our motor skills effectively, to control our gestures and direct our movements, seems self-evident. Yet many collisions and falls are linked to a lack of mastery of this coordination between our vision and the movement of our limbs.

Rules and injunctions such as the ban on running in the workplace, the obligation to hold on to handrails when climbing up and down the stairs, or simply looking where you go and what you do will be respected all the more by everyone if these sensorimotor mechanisms are clearly explained and understood.

Conditioned Reflexes

Unlike innate reflexes, such as the babies' grasping reflexes, **conditioned reflexes** (also known as conditional or acquired reflexes) develop with experience. Such experience establishes a connection between a trigger stimulus, a 1 and a consequence. When a behaviour allows us to succeed systematically in a daily situation, it becomes automatic.

We have thus developed the reflex of trying to catch falling objects to prevent them from being damaged, dodging obstacles in our path, or protecting our faces when we see a projectile coming our way.

All this is generally useful and helpful, and that is why these reflexes are acquired and perpetuated. But in certain circumstances they can also do us a disservice and cause accidents. Let us consider again the avoidance reflex: a driver will find it very difficult to stop herself from swerving when an obstacle suddenly appears in her path, even if this results in going off the road with more serious consequences than the impact with the obstacle in question.

FOCUS ON... THE ORIGINS OF BEHAVIOURISM

At the end of the 19th century, Ivan Pavlov made his discoveries about conditional reflexes (Pavlov, 1935), preceding the work of John Watson, who in 1913 established the principles of behaviourism (Watson, 1913), followed by B.F. Skinner in 1938 with his work on operating behaviours (Skinner, 1938; McLeod, 2015).

RESPONDENT BEHAVIOUR

Like the economists of the neoclassical school at the same period, Watson also aimed to elevate the study of behaviour to the level of Natural Sciences. Psychology was therefore to disregard mental states (more on those later) and focus solely on observable phenomena.

He devised and carried out laboratory experiments. The science of behaviour as he understood it thus became the study of the relationships between the stimuli perceived by an individual and the behaviours that this individual performs in response, relationships that can be measured.

Stimulus → Individual → Response

In this first version of behaviourism, since the individual is considered to be a black box, all these internal phenomena are simply ignored, and the schema is simplified as follows:

Stimulus → Response

OPERATING BEHAVIOURS

But the above approach only explains part of our behaviours. Skinner therefore took up this scheme and introduced into it the consequences that the individual will experience as a result of the behaviour (response) that was performed.

His experiments with animals enabled him to prove that their behaviour was more influenced by the consequences than by the antecedents of that behaviour.

The principle therefore became:

Stimulus → Response → Consequence

This process is no longer linear but a learning loop. It is the consequences of a behaviour (favourable or unfavourable) that will condition the behaviour that will follow in response to a stimulus with which it would have been associated.

REINFORCEMENT AND PUNISHMENT

Consequences are categorized according to the effect they have on the likelihood of the preceding behaviour being repeated in the future. A consequence is said to be *reinforcing* if it increases the frequency of recurrence of the behaviour that precedes it, and *punishing* if it decreases it.

Both reinforcement and punishment can be positive or negative. This positive or negative character should not be understood as favourable or unfavourable, but in the "mathematical" sense of the term: positive meaning that a stimulus is added or increased, and negative meaning that it is subtracted or decreased.

This gives us the four possible combinations below:

- **Positive reinforcement**: the individual obtains something he or she appreciates.
- **Negative reinforcement**: the individual is relieved of a constraint.
- **Positive punishment**: the individual acquires a constraint.
- **Negative punishment**: the individual loses something they value.

The application of behaviourism to our affairs, known as *behaviour-based safety* (BBS), will be presented in our Toolbox (Tool No. 4).

Key Learning Points for the Safety Leader

Once acquired and consolidated, it is almost impossible to control one's reflexes. Only a handful of elite teams, typically in organizations facing extreme risks that cannot be otherwise mitigated, face such a challenge (like the US Navy's Blue Angels and the RAF's Red Arrows). And they deploy selection and training schemes that cannot be realistically implemented by most civil organizations.

In the occupational safety field, our reflexes and their role in the various activities (often serving us and sometimes disserving us) must be identified and treated for in risk assessments.

Procedural Automatisms

Leave the house, take a few steps, and then ask oneself: "Did I lock the door? Turn around, grab the handle, and see that the door is locked. It has happened to all of us. How is it possible that we have no memory of a series of actions (retrieving and grabbing the key ring, finding and inserting the right key in the lock, turning it, taking it out again, and putting the key ring back in our pocket or purse...) that we performed just a few seconds ago?

We automate sequences and routines, or **procedures** as psychologists call them, that always run identically (Figure 42). This ability is essential to our day-to-day functioning. Typing your password to launch your laptop, putting your cereal back in the kitchen cabinet after breakfast, lacing your shoes... tasks that once involved a conscious process, and which initially required our full attention, are carried out automatically once managed by our *procedural memory*, and we can even do something else at the same time, such as chatting or singing (although this is not always a good idea, see the topic on divided attention above).

The same is true in occupational settings for gestures ranging from starting the sequence of a machine to opening an oven, taking the baked item out of it, putting

FIGURE 42   Routines (drawing by author).

a raw item in, and then closing it again. But there are risks involved. Operators who always perform the same gestures in the same order will need to focus their full attention to following a new protocol, which introduces a variation in one of the steps; the slightest drop in their level of concentration (and we have seen that our concentration is far from perfect) will cause them to reproduce the order they had internalized. Even without a change in the routine, if a distraction causes them to lose the thread of the sequence, they can easily miss, double, or swap steps.

DID YOU KNOW? AUTOPILOT

Car manufacturers are working hard to bring us autonomous cars. In fact, we are already driving on "autopilot" in our daily lives and have been for a long time!

The combination of the three phenomena described above and the repertoire of automatic gestures and reactions that result from them means that, when on the road, most of the time we do not consciously analyse and control our actions:

1. Our **sensorimotor intelligence** enables us to follow and maintain a trajectory, in particular with the help of lines drawn on the road and by following the vehicle ahead. It also optimizes the collection of useful information by directing our gaze and, on the basis of the information collected (for example determining whether another driver has seen me or not), enable us to circulate among other vehicles by dynamically calculating their probable trajectories as a function of countless parameters.
2. Our **acquired reflexes** make us slow down when we see a red light ahead, brake immediately when we see the braking lights of the car in front of us come on, turn the steering wheel to avoid an obstacle, etc....

3. Our **procedural memory** allows us to shift gears without thinking about it and get to work every morning without any need to reflect on our way (to the point that we are sometimes surprised that we've "already arrived" because our minds were busy considering other things, like the morning meeting with our boss).

Once again, all these mechanisms work very well in general, i.e. in the absence of any situation that is out of the ordinary. As a result, we almost inevitably do not focus our conscious attention on driving when we know well the route and under normal circumstances. It is only a small step from there to thinking that we can drive and do something else at the same time (like reading a message that has just arrived on our phone), a step that many drivers thoughtlessly take every day, with the consequences that we know.

The same mechanisms, with the same advantages, the same limitations, and the same unfortunate consequences, operate when we are carrying out routine activities at work!

Key Learning Points for the Safety Leader

The automatisms that we develop through professional experience are essential to our working performance. Our ability to recognize cues and patterns (which we will study latter) and react swiftly allows us to be efficient, as these reactions are generally adequate, fluid and precise. They save us time and effort; they free our working memory, and their progressive acquisition sets apart the skilful worker from the beginner.

But automatisms are also a direct cause of many accidents, because we reproduce a behaviour without any conscious supervision on our part, while feeling that we are in control. Critical tasks in which such automatisms are at play or may appear must be designed so as to incorporate specific protocols, technological safety assistance or external supervision.

Several of such means are presented in our Toolbox, Tool No. 3.

LONG-TERM DECLARATIVE MEMORY

The storage of explicit information over the long term is managed by two closely intertwined systems:

- Our **episodic memory** stores the moments experienced by the individual (memories), constituting our spatiotemporal references and enabling us to project ourselves into the future.
- Our **semantic memory** manages language and knowledge about the world and ourselves, regardless of the context in which this information is acquired.

The brain does not store this information in the same way as the read-only memory of a computer. Both the process of consolidating information and that of retrieving it are carried out by neurochemical reactions and facilitated or inhibited by physiological, psychological, and situational factors.

Consolidation

It is the **consolidation** by the hippocampus of the data acquired by our perceptual memory that feeds our long-term memory, their use by our working memory facilitating this process. Synapses that are sensitized more or less temporarily when "short-term" memory is created can be strengthened and become durable if the same signal repeatedly reactivates them, a phenomenon called long-term potentiation. In turn, and as mentioned above, the information already stored in long-term memory constitutes antecedents that will influence what is retained in perceptual memory and how it is handled by the working memory.

In addition, certain information is more or less effectively consolidated and therefore more or less permanent and easily accessible at a later date: the hippocampus is influenced by the amygdala, which reinforces the consolidation of information linked to emotions. Moderate stress related to the task can therefore promote the consolidation of memories, as the amygdala and hippocampus are stimulated by corticosteroids; however, extreme or chronic stress ends up reducing the capacity to memorize new information.

Practical learning (in a controlled environment) aimed at developing the mastery of risky operations will always be more effective than purely theoretical learning because of the involvement of episodic memory and the moderate – and therefore positive – stress associated with the situation. Excessive stress, however, is detrimental to learning.

To facilitate the consolidation process, individuals can use **mnemonic techniques**. There are many different methods, but they are all based essentially on three principles (Simonetto, 2020):

- *Structuring information* by category, theme, chronological order, colour, etc.
- *Association of ideas,* thus "hooking" new information onto already consolidated information.
- *Visualization,* mentally representing the scene in detail.

As well as acquiring these strategies, which are eventually implemented spontaneously, workers who exercise their memory on a daily basis develop their hippocampus, which increases in size.

DID YOU KNOW? ON TAXI DRIVERS' BRAIN SIZE

Taxi drivers need a good memory. In London, they have to remember 25,000 streets within a 10 km (6 mi) radius of Charing Cross as well as thousands of tourist attractions and other points of interest. They spend 3–4 years as trainees, after which only about half of them pass the exam to drive a "black taxicab".

Neuroscientists followed 79 of these trainees over a period of 4 years and studied their brains using MRI, comparing them with those of a similarly educated population who had not been trained as taxis. At the outset, the hippocampus of all the subjects was of equivalent size and they all achieved comparable scores on memory tests.

At the end of the 4-year training period, the hippocampus of the 39 trainees who had passed their exams was significantly larger than that of the others (Woollett and Maguire, 2011).

Key Learning Points for the Safety Leader

Organizations carrying out activities whose safety depends on operators memorizing information must:

- structure critical information to make it easier to remember,
- train operators in the most appropriate memorization strategies,
- where necessary, plan, design, and implement tools and support for externalizing knowledge storage, as described below.

Retrieval

Even moderate stress hinders the active **retrieval** of information from our long-term memory: the stimulation of the amygdala by corticosteroids (responsible for triggering the primary automatic fight–flight reactions mentioned above) inhibits the prefrontal cortex in a stressful situation, and even more so if it is perceived as dangerous. When in panic mode, we cannot remember even the most familiar information.

Beyond these stressful situations, the most typical recovery problems are as follows:

- **Interference**: the structuring of information and associations of ideas mentioned above can lead to retrieval errors. Routine situations subconsciously trigger nerve impulses that activate associated information. We think we have seen or done what we would normally have seen or done in the same context.
- **Approximate reconstruction**: the limits of our attentional resources and the resulting selectivity (as described above) mean that a great deal of information is simply not effectively recorded in our long-term memory because it has not been processed by our working memory. So when we try to retrieve memories of a given episode, we subconsciously "fill in the gaps" with data that seem plausible to us.
- **False memories**: this same reconstruction mechanism can go as far as implanting the memory of a completely fictitious episode. Information never acquired by the subject but which is insistently presented to her as having

been experienced by her ends up being incorporated into her memory, and is often even enriched by her with other elements by the association mechanism that we saw above (Corson and Verrier, 2013).

- **Lapses of memory**: the technical, technological, and organizational complexity inherent to many tasks and jobs means that individuals will not be able to memorize and retrieve effectively all the information they may need, especially if it is not used periodically – and therefore not "refreshed" in their memory.

The typical solution is to externalize the storage of information in the form of additional expertise within a team, reference documents, or technological aids. These external supports are often necessary, but they reduce the individual's motivation to make the effort required to memorize information ("since I've got a GPS, I can't remember the routes, so I always need the GPS"), which can pose a problem when these external resources are not available.

Key Learning Points for the Safety Leader

Protocols and other emergency response procedures must anticipate and take into account the difficulty operators may have in remembering information in stressful situations.

More generally, it is in every organization's interest to put in place systems to ensure that the operator has access to the information needed to carry out dangerous tasks, and which could be forgotten even by experienced workers, at the place and time they are carried out. These devices can be documented instructions (on paper or screens), signs, warning lights, sound signals, etc.

The most suitable tools and media will have to be chosen, taking into account the limits of perception and attention outlined above, as well as the motivation of workers to seek out and use artefacts that they may consider redundant or useless in the light of their experience. We will return to this motivational dimension later.

TOOLBOX – Tool No. 3: Attention Management Techniques

This tool is a compilation of several short descriptions of tools more or less well-known to facilitate attention management, as well as to adapt the job to the limits of our attentional resources (Blanco-Munoz, 2021).

FAIL-SAFE DESIGN

Fail-safe design aims at integrating built-in safety features in the objects, tools, and, in general, any interface that a worker will need to interact with, and which could lead to accidental damage if used wrong. Also known by the Japanese expression *poka-yoke*, or more prosaically as "*foolproof design*" (a term that most human factors' specialists consider abusive, and rightly so), they encompass a wide variety of techniques that make it mechanically impossible to mistakenly introduce or remove, connect or disconnect, plug or unplug, assemble or disassemble, switch on or off... parts or systems.

This principle can be applied not only to component design but also to software design (for example making it impossible to validate a step without having chosen a value in a given field) and, more generally, to the design of any work interface, physical or virtual, involving choices that can be effectively constrained and whose errors can be anticipated a priori.

Examples from everyday life are SIM cards in mobile phones or SD memory cards in cameras, which, like the old 3.5-inch floppy disks, have a notch to prevent them from being inserted the wrong way.

5S

Visual management can be considered a derivative of choice architecture or *nudge* even if its deployment in industrial environments preceded the conceptualization of that idea.

This tool places the emphasis firmly on visual information because, unlike reading, which involves cognitive processes that take place at a high cortical level and require more effort, our evolution has allocated huge areas of our brain to vision processing. We might as well align our work environments with our perceptual and attentional capacities.

A very popular and effective technique for structuring this principle is the **5S**. It became widely known as Toyota integrated it in its famous TPS (*Toyota Production System*), and consists of five simple steps whose name in Japanese start with the letter S, and that were translated in English by words more or less equivalent and starting with the same letter: *Seiri* (sort), *Seiton* (set), *Seiso* (shine), *Seiketsu* (standardize), and *Shitsuke* (sustain).

The so-called "5S projects" seek to optimize visual perception and organize the information made available to the operator so as to draw her attention to hazards and, in general, to any object she needs to spot, as well as to direct her movements and gestures without constraining them.

By eliminating everything that is unnecessary, organizing the workspace efficiently, improving the cleanliness of the premises, and systematizing practices, we not only eliminate latent conditions that could contribute to the occurrence of an incident, but we also simplify the working environment, making it more compatible with the operator's attentional resources, and thus limiting the probability of error.

The organization of elements (tools, parts, material, equipment, waste, etc.) in the workers' visual environment enables them to notice in particular the elements that have changed (been modified, moved, appeared, or disappeared).

The main difference with the good old "order & cleanliness" injunctions is the emphasis placed on the fifth and last stage: the formalization of the new norm, which must absolutely be respected by all users of the space in question; otherwise, the mess will quickly return. This works very well in Japan, where everyone is imbued with a culture of order and rigour, particularly when it comes to cleanliness and tidiness, with pupils being responsible for cleaning and tidying classrooms and canteens at school from a very early age.

We will come back later to the key role that the acquisition of these values plays in behaviour.

SECURE READING

If you know Dr. Kawashima's "brain training programme", available on a very popular games console, you may have tried the "Stroop Test". The game displays one of four words and colours: red, blue, yellow, and black. One of these words appears on the screen, its font in a random colour that may or may not match the colour denoted by the word. The challenge is to say out loud the font's colour, and not to read the word. It is more difficult than you may think!

In fact, reading for hours a day since we were children has become so automatic that we cannot help reading a word or text that is put in front of us. But this reading can be totally automatic, without paying any attention to it: like when we are reading a book or a document and, in the middle of a sentence, we realize that we were thinking about something else and that we have not copied anything we have just read.

This also happens at work, especially when the operator must refer to the same written instruction or checklist regularly. **Secure reading** using sensory triangulation reduces operator errors when taking instructions from a written document.

This technique simply consists of reading half-aloud and following the text with your finger. This activates not only vision but also hearing (you can hear yourself), touch, and motor skills. Engaging these different areas of the brain results in a shift to a more conscious, analytical mode (Simonetto, 2020).

MENTALLY PREPARING FOR THE JOB

Research into the brains of sportspeople using different mental imagery techniques has shown that simply visualizing motor activity activates the same areas of the brain as actually performing the activity (Robin and Flochlay, 2017). Alpine ski champions mentally replay their runs before they take to the slopes, achieving the same time in their heads (to within a few tenths of a second) as they did in training.

Perhaps the best-known application of these visualization exercises is the rituals used by elite fighter pilots (the Blue Angels or Red Arrows we mentioned before) as part of their training, particularly before flying in close formation.

Other organizations at the forefront of human factors have incorporated mental preparation for operators before carrying out high-risk tasks. The **Pre-Job Briefing** used in the nuclear industry aims to exploit these visualization/execution links to mentally prepare operators, particularly for dealing with the unexpected (EDF, 2022).

This same technique of mental preparation for the task can also be deployed in the context of occupational safety. In the automotive industry, for example and particularly at Toyota, operators on certain assembly lines start their shift with motor skills exercises. These exercises not only warm up and stretch their muscles but also help them rehearse the movements they will have to perform on the line, thereby pre-activating the neural networks that will come into play, as in the *priming* presented earlier.

DOWNTIME – PAUSE BEFORE ACTION

A series of fairly universal good practices that all parents try to instil in their children revolve around simple concepts such as think before you speak, look before you cross, check the depth of the pool before you dive... If it is not easy to get children to adhere to these practices, it is because they do not come naturally: day-to-day experience tends to reward certain lapses in thinking and other risk-taking that save us time, with serious accidents remaining rare...

It is the same at work: pausing briefly before carrying out an action is a small cost compared to the potential loss that would result from the accident, but it is a price we are generally not willing to pay, because of mechanisms that we will explain in detail in the next chapter.

It is very difficult, if not impossible, to just impose this kind of behaviour on an individual, let alone on a group within an organization. But we can nevertheless mobilize our efforts to establish downtime as a practice that can be ritualized as part of a cultural evolution process, using the techniques that will be described further in this book.

Some organizations have successfully introduced practices such as "**Take 2**", which encourages employees to systematically take:

- *2 seconds* to look carefully before making each gesture,
- *2 minutes* and *two steps back* to look 360°, up and down, before starting any task, even the most routine ones (and perhaps especially those, as we tend to carry them out without thinking),
- a *second opinion* from a colleague or supervisor if our first analysis reveals the slightest anomaly or raises the slightest doubt,
- *2 hours*, or as long as it takes, to discuss the issue if necessary and assess the risks in detail, calling on additional expertise if necessary.

Tool No. 9 presents a more sophisticated variant of this downtime called dynamic risk assessment. These two tools complement and reinforce each other as safety practices.

To promote this type of tool and the mindset that underpins it, there are also organizations that run mindfulness programmes, which can have beneficial effects on a wide range of objectives, from mental health and well-being at work to quality and safety.

But let us say it again: we cannot expect these practices to take root and flourish if the organization's culture does not support them.

SHISA KANKO: POINTING AND CALLING

In Japan we can see railway workers in the platforms and runway operators in the airports perform a very interesting ritual called *shisa kanko* (or shisa kakunin kanko). Initiated in the early 1900s by train drivers, it is now widely used in Japanese industry.

It very simply translates as "pointing and calling", and it is as simple as that: gesturing and verbalizing the status of the different indicators they need to check, whether a switch in a control panel or the perfect sealing of a door.

At the beginning of the last century, steam locomotives were extremely loud and generated a lot of steam and smoke. These conditions complicated the cooperation between the train drivers and the other team members, so they had to call out loudly. To this they integrated the pointing a few decades later.

They realized that this combination of verbal and body gesture confirmation reduced the error rate significantly. As for the secured reading presented above, the fact of mobilizing several regions of the brain, including the visual and auditory cortices, the motor cortex, and Broca's area (involved in speech), ensures that the working memory is fully focused on the task.

This way of operating requires the individuals to deploy their complete attention and prevents them from being distracted and from mechanically going through the motions while having their attention focused on something else.

ERROR TRAINING

Learning a skill is traditionally based on passing on the "right" knowledge and repeating the "right" gestures and actions. However, it has been demonstrated that training in which students are not only encouraged to experiment but also led to make mistakes results in better learning in the short- and medium-term. First-hand experience of the failure of a technique, tactic, or strategy is beneficial on several levels:

- The individual is forced to consider the possibility that everything may not always go according to plan.
- They realize the complexity of the system they form with the task, the tools, the colleagues, the materials and components, the equipment and the environment, and the multiple interactions between all these elements, which are difficult to predict and even more difficult to explain in procedures or even in direct verbal instructions, which are always subject to interpretation.
- They realize the limits of their own logic, limits that we will review in detail later.
- Perhaps most importantly, we mitigate the tendency to overconfidence, which will also be discussed when we talk about cognitive biases.

The only condition is that these exercises also include instructions on how to deal with the error that has been made (Heimbeck et al., 2003).

It goes without saying that the experience of the unexpected and of error in the context of training cannot involve a risk of physical or psychological harm to the worker or to a third party. But simulation exercises can lead the individual towards the wrong choice or action and then show them the consequences that these would have caused without necessarily triggering them. New immersive teaching tools in virtual and augmented reality environments can take this approach even further.

TOOLBOX – Tool No. 4: Behaviour-Based Safety

The processes of learning by association, which were demonstrated experimentally by Pavlov, Watson, and Skinner, and developed at a neurobiological level by Hebb, occur naturally when an individual interacts with their environment. They also appeared to be a model that can be used as part of a strategy to promote or eradicate targeted behaviours (Blanco-Munoz, 2019).

The first projects trying to apply this approach to occupational safety took place in the United States in the late 1970s (Komaki et al., 1978; Krause et al., 1996). They were at the origin of **BBS**, which spread rapidly, initially in English-speaking countries and then throughout the world, as large multinationals sought to roll out these programmes across all their sites and subsidiaries.

Skinner's stimulus–response–consequence model is used again, but it is often renamed ABC for antecedent (or activator)–behaviour–consequence.

The starting point for these programmes is that workers take risks because these "risky behaviours", even if potentially harmful, are most of the time reinforced, whether positively (for example setting up a machine without locking it out saves time) or negatively (for example not wearing protective gloves relieves a feeling of discomfort). The aim of BBS programmes is to reverse this logic, implementing initiatives to ensure that so-called risky behaviour is punished (in the behaviourist sense of the term) and preventive behaviour is rewarded.

BEHAVIOUR OBSERVATION SCHEMES

The main tool used in these BBS programmes is observation schemes. Observers, generally members of the organization's staff and therefore with a good knowledge of its activities, are trained to identify target behaviours (or safety-critical behaviours, determined by a statistical study of accidentology and by analysis of major risks), then observation routines are implemented.

The frequency, duration, and coverage rate of these observation sequences must be sufficient to ensure that the target behaviours have a high probability of being detected and therefore of being subject to the appropriate consequences.

This is the cornerstone of this approach: observers must provide workers with the appropriate consequences in the form of constructive feedback.

DELIVERING CONSTRUCTIVE FEEDBACK: PIC-NIC

PIC-NIC is a mnemonic acronym that summarizes the characteristics of effective feedback from an observer to an observed individual.

For these consequences to have an influence on the probability of recurrence of the behaviour in question, they must be:

$$\text{Positive} + \text{Immediate} + \text{Certain (Constant, Coherent)} = \textbf{PIC} \text{ or}$$
$$\text{Negative} + \text{Immediate} + \text{Certain (Constant, Coherent)} = \textbf{NIC}.$$

Positive or **Negative**: Appropriate behaviours must be recognized, and the individual carrying them out must obtain positive consequences (reinforcement). Conversely, unsafe behaviours must be followed by unfavourable consequences known as "negative" (punishment). It should be noted that, to obtain a catchier acronym, this terminology differs from that used by Skinner and described previously. For the purposes of communication, it is also interesting to avoid the term punishment, which in everyday language implies a disciplinary, moralizing, or even paternalistic connotation.

Immediate: Whether "positive" (reinforcement) or "negative" (punishment), the feedback must follow the behaviour as quickly as possible. And not only to make the dangerous situation cease: it has been demonstrated that the quicker the consequence follows the behaviour, the stronger the association and thus the more effective the learning experience (Epstein et al., 2002). Moreover, delayed feedback is not only less effective in terms of learning, but it also tends to demonstrate that the observer does not consider the behaviour to be sufficiently important to deserve instantaneous feedback, which undermines the construction of prevention as a value (a subject we will return to later).

Certain (Constant, Coherent): Whether "positive" (reinforcement) or "negative" (punishment), as well as being immediate, feedback must always be the same. Obviously, the individual will not get any feedback in the absence of an observer, but as soon as a worker is observed performing a behaviour, the observer must systematically provide them with feedback which must be consistent with the stated objectives of the approach and coherent over time, regardless of the circumstances (production rushes, emergencies, etc.).

OBSERVATION SCHEMES' SUCCESS FACTORS

Here are listed the main keys to successfully implementing these observation schemes:

Organization and preparation

1. Select a sufficient number of motivated observers from among the staff and train them to:
   i.   identify the target behaviours and
   ii. deliver constructive feedback (PIC-NIC).
2. Define the frequency with which each worker must be observed and draw up a programme of observation sequences, defining the periodicity and duration accordingly.

3. A typical observation session with a team, for example a dozen employees in a $300\,m^2$ ($360\,yd^2$) workshop can last between 15 and 20 minutes.
4. Observers should have no hierarchical link with the individuals being observed. Ideally, cross-observations should be made between teams or departments.
5. Pre-printed observation cards can be provided to make it easier to record observations. Mobile applications are also available on the market for this purpose.
6. Before it is launched, regulations and company agreements may require the programme to be presented to the company's staff representatives, safety committees, and/or unions. In any case, its objectives and procedures must be widely communicated to all employees, including temporary staff and contractors.

Execution of the observation sessions

1. The observers must introduce themselves to the individuals they are about to observe and tell them the purpose of their visit.
2. The observer should not be looking only for inadequate behaviour but should be just as keen to identify appropriate behaviour.
3. The main objective of the observer is to identify the target behaviours and provide the corresponding constructive feedback.
4. If notes are taken, they will only concern information relating to the behaviour and its context (environmental, technical, and organizational details). Under no circumstances will the observer record the identity of the individual(s) observed.
5. The observer will not film or take any photographs of the individual being observed or of the task in progress without the individual's consent.

Escalation and consolidation of results

1. Qualitative information from observation sequences must be debriefed with the manager of the team observed as soon as possible.
2. The identity of the workers observed should not be passed on by the observer to the team leader, only the nature of the behaviour observed and the discussions held with the individuals.
3. The data from the information sequences can be consolidated for statistical analysis. Ad hoc software is available on the market.
4. The quantitative and qualitative information gathered and the conclusions drawn from its analysis must be used to identify, prioritize, and resolve the technological and organizational problems that constitute the latent conditions, the real root causes of accidents. Many organizations (and consultants) forget this last, yet crucial, step (Krause and Bell, 2015).

EXAMPLE OF OBSERVATION TEMPLATE

The grid below is an example developed and implemented by the author, which, compared to other templates, has the advantage of:

1. Proposing a dozen "direct causes" linked to human factors, some of which have already been presented and others which will be presented later, so that the observer can discuss them with the worker being observed. This requires the observers to be trained in human factors and the observed workers to be provided with at least a basic awareness on the topic.
2. Providing the observer with guidance to assess the behaviour observed: moderate risk, serious risk, and safe behaviour – because safe behaviour must also be spotted and recognized.
3. In the event of unsafe behaviour, it proposes a section serving as a basis for the immediate feedback to the worker in the form of a discussion on the three prerequisites for adopting safe behaviours (Three Traffic Lights model or "I should/I can/I do"), which will be presented at the end of this work.
4. If the observed worker disagrees with one of the three prerequisites, the grid presents a final section to be used as a basis for debriefing with the worker's manager on the nature of the difficulties – without giving the manager the identity of the worker who expressed them unless authorized by the worker to do so (Figure 43).

BBS'S LIMITS AND PITFALLS

For over 40 years, thousands of programmes based on the principles of BBS have been deployed in a wide variety of organizations. Numerous authors and, above all, a multitude of consultancy firms have seized upon these theories and "packaged" them in a wide variety of formats.

As we shall see in the next chapters, from a theoretical point of view, BBS has been challenged by the emergence of new theories questioning the capacity of behaviourist approaches to explain and influence the cognitive, affective and social dimensions of behaviour.

In practice, the following difficulties have been encountered when rolling out these programmes:

1. **Cultural incompatibility**: Originating in the English-speaking world, BBS has not always exported well to other regions. Particularly in Latin cultures, its development has undoubtedly been hampered by the prevalence of a significant power distance and an equally strong intolerance of uncertainty. These cultural factors will be discussed later.

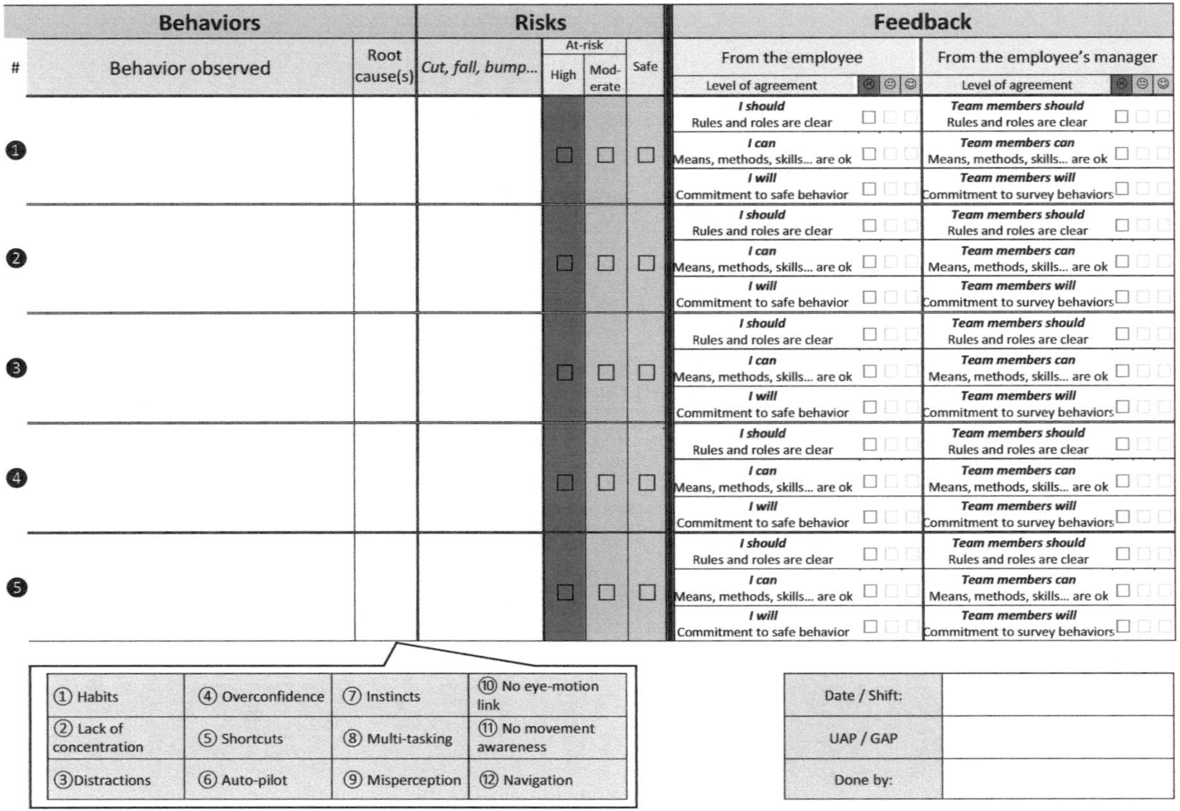

	Behaviors		Risks				Feedback	
							From the employee	From the employee's manager
#	Behavior observed	Root cause(s)	Cut, fall, bump…	At-risk High	At-risk Mod-erate	Safe	Level of agreement ☺ ☺ ☺	Level of agreement ☺ ☺ ☺
❶				☐	☐	☐	*I should* Rules and roles are clear ☐ *I can* Means, methods, skills… are ok ☐ *I will* Commitment to safe behavior ☐	*Team members should* Rules and roles are clear ☐ *Team members can* Means, methods, skills… are ok ☐ *Team members will* Commitment to survey behaviors ☐
❷				☐	☐	☐	*I should* Rules and roles are clear ☐ *I can* Means, methods, skills… are ok ☐ *I will* Commitment to safe behavior ☐	*Team members should* Rules and roles are clear ☐ *Team members can* Means, methods, skills… are ok ☐ *Team members will* Commitment to survey behaviors ☐
❸				☐	☐	☐	*I should* Rules and roles are clear ☐ *I can* Means, methods, skills… are ok ☐ *I will* Commitment to safe behavior ☐	*Team members should* Rules and roles are clear ☐ *Team members can* Means, methods, skills… are ok ☐ *Team members will* Commitment to survey behaviors ☐
❹				☐	☐	☐	*I should* Rules and roles are clear ☐ *I can* Means, methods, skills… are ok ☐ *I will* Commitment to safe behavior ☐	*Team members should* Rules and roles are clear ☐ *Team members can* Means, methods, skills… are ok ☐ *Team members will* Commitment to survey behaviors ☐
❺				☐	☐	☐	*I should* Rules and roles are clear ☐ *I can* Means, methods, skills… are ok ☐ *I will* Commitment to safe behavior ☐	*Team members should* Rules and roles are clear ☐ *Team members can* Means, methods, skills… are ok ☐ *Team members will* Commitment to survey behaviors ☐

① Habits	④ Overconfidence	⑦ Instincts	⑩ No eye-motion link
② Lack of concentration	⑤ Shortcuts	⑧ Multi-tasking	⑪ No movement awareness
③ Distractions	⑥ Auto-pilot	⑨ Misperception	⑫ Navigation

Date / Shift:	
UAP / GAP	
Done by:	

FIGURE 43 Behaviours observation grid (drawing by author).

2. **Regulators' mistrust**: As previously mentioned, Directive 89/391/
   EEC anchored in the European Union, the consensus established
   among these countries on the primary responsibility of the employer
   when it comes to occupational health and safety. It places at the top
   of the nine "general principles of prevention" the assessment and prevention of risks upstream and highlights the use of technical means
   and the organization of work. Only lastly does it deal with the training
   and information of workers and therefore the question of behaviour
   as addressed by BBS. Approaches based on behaviour are thus perceived by the legislators transposing this directive and by the national
   authorities and agencies responsible for ensuring compliance with it
   as subsidiary at best, and at worst as a means for the employer to shirk
   their responsibilities.
3. **Intensity and coverage of observation routines**: Given the number of behaviours performed by each worker on a daily basis (several thousands), it has often been argued that the schemes typically
   deployed do not achieve the number of observations and feedback
   required to have a real impact. Indeed, to generate antecedents
   capable of effectively and sustainably influencing workers' attitudes
   (a concept that we will develop later), the feedback they receive must
   be continuous and not limited to observation visits.
4. **Dispersed teams and isolated workers**: Observation schemes are
   very difficult to deploy in organizations with teams dispersed over
   different sites, and even more so with regard to isolated or itinerant
   employees.
5. **Tendency to focus on punishment**: The prevalence of the neoclassical economic view of behaviour (a theory to which we shall return),
   which implies that individuals are not supposed to make mistakes,
   means that it is much more natural for an observer to react to a deviation than to recognize and highlight appropriate behaviour, which
   is perceived as "normal". If observers are not trained to avoid this
   bias and "negative" feedback (punishment) predominates, prevention
   itself ends being perceived negatively by the workers.
6. **Hawthorne effect**: The simple fact of knowing that they are being
   observed alters the individuals' behaviour, who tend to adopt responses
   that are more in line with those expected than those they would have
   made spontaneously in the absence of an observer. Once the observations have ceased, the individuals revert to their usual behaviour.
   This often-unconscious phenomenon, known as the Hawthorne
   effect, which we will describe in more detail later (Gillespie, 1991),
   may call into question the approach and, above all, the relevance of
   using quantitative data as a leading indicator of safety (see point 7
   below).

7. **Placing too much emphasis on derived indicators**: Despite the fact that the number of behaviours observed may not be statistically significant (point 3 above) and that the observations may be biased by the Hawthorne effect (point 6 above), it is common for organizations deploying BBS observation schemes to use the data collected to prepare indicators. While monitoring the total number of observations done may make sense to ensure the continuity of the approach (point 8 below), calculating ratios of unsafe behaviour observed does not seem relevant as a leading indicator of prevention.

We will address the matter of measuring safety, a very slippery topic, in a dedicated section.

8. **Process running out of steam**: Most organizations that set up observation schemes find that, after a while, observations and interesting feedback become increasingly rare. As a result, almost all organizations end up stopping using them, at least in a formal way.

The optimistic view is that the approach works: unsafe behaviour becomes rare, and we lose the motivation to congratulate preventive behaviour which becomes obvious, even automatic. Ideally, giving feedback would also become natural, and you would no longer need to be in an observation sequence to do so.

The pessimistic view is that the players become weary, because the whole organization is not evolving towards a culture of prevention as presented in the following sections.

9. **Targeting only the workers on the front line**: Finally, the main criticism that can be levelled at observation approaches is that they have traditionally focused on operators in the field. They may therefore have the impression that the organization is directing all its efforts towards preventing unsafe behaviour at the end of the pipe, instead of also investing in strengthening the other elements of the system, whether this be engineering solutions by reviewing design choices and equipment's obsolescence, or system improvement by seeking to elevate management practices.

It was the recognition of these shortcomings and weaknesses that led to the development of "new approaches", even if in absolute terms we cannot deny the usefulness of BBS as a tool that can be deployed as part of the cultural change process that we will describe.

3 The Mind
Rationality, Attitudes, and Motivation

"There can be no knowledge without emotion. We may be aware of a truth, yet until we have felt its force, it is not ours".

Arnold Bennett

Understanding the physiological, sensory, attentional, and mnemonic limits described in the previous sections enables us to gain a better understanding of the capabilities of the human factor and to identify an initial series of weaknesses, but it is also necessary to study the limits of our **rationality**.

In fact, our *deliberative processes* are built on the basis of a representation of the situation that emerges from both the information that we perceive and the information accessible in our memory (with varying degrees of accuracy in the case of both the former and the latter, as we have just seen). We will now study how these elements are processed in our minds.

The individual's capacity for reflection, judgement, and engagement is obviously invaluable when faced with tasks and work environments that are always more complex than they appear. But these faculties, like those we have seen so far, also have their limits, which we need to understand to find the best socio-technical means of compensating for them.

RULE-BASED SAFETY VS MANAGED SAFETY

"I love the rule that corrects the emotion. I love the emotion that corrects the rule".

Georges Braque (1882–1963)

As we all know, it is not enough to decree that a task must be carried out and how it must be done to ensure that the work is performed in full compliance with the specifications. And that is a good thing because following instructions with absolute zeal would often result in failure (Blanco-Munoz, 2019).

Conflicts between *prescribed tasks* (also called work-as-imagined) and *actual work* (or work-as-done) have been studied since the 1950s. The work of Ombredane and Favergé (1955), Leplat and Cuny (1977), and Leplat and Hoc (1983) is still topical, particularly through Hollnagel's high-profile work on Safety-II (Hollnagel, 2014), addressing the differences between what he calls *tractable systems* and *intractable systems*.

DOI: 10.1201/9781003617341-3

FOCUS ON... TRACTABLE SYSTEMS VS INTRACTABLE SYSTEMS

Tractable systems are characterized by linear relationships, governed by easily identifiable causal links between components that can be isolated and described individually and whose operation is bimodal (they either operate as expected or they fail, with possible well-defined "degraded modes" in between). The operation and results produced by these systems, which can be represented as processes in the technological sense of the term (input–processing–output), are therefore predictable.

Intractable systems, however, correspond to the definition of complex adaptive systems (Holland, 1992) and are characterized by the attributes issued from complexity theory (Bar-Yam, 2002). The elements that make them up are all more or less *interdependent*, and the relationships between them are non-linear: they are not necessarily direct or close in time or space, they are not necessarily proportional, and they give rise to *emergent behaviours*, such as the ability to self-organize, to anticipate, and to evolve, which operate with a very high degree of variability and whose explanation eludes analyses seeking to establish direct causal links.

The variability of these intractable systems is not purely stochastic but the result of the adjustments that each of the elements makes in relation to the others within the system: adjustments that are both *reactive*, depending on the behaviour of the other elements, and *proactive*, depending on their own needs and expectations.

REGULATING STABLE SYSTEMS

In very stable, predictable systems that lend themselves to a high degree of automation, such as production lines and other highly mechanized or even robotized workplaces, it is possible to control a large proportion of activities in a purely technical way, thereby significantly reducing the probability of errors and accidents.

These are generally industries impregnated with a process culture, where engineers are held in high regard and where the tasks that still require human intervention are often approached using the same "mechanistic" approach. From an organizational point of view, we might also consider applying the logic of rules and control to these systems. It might be tempting to think that operators do not have to think about what they are doing.

DID YOU KNOW? TAYLORISM AND FORDISM

Although specialization and division of labour had already been theorized by Adam Smith in the 18th century (Smith, 1776), it was by the end of the 19th century that Frederick W. Taylor proposed the principles of **scientific management** of work (Taylor, 1911) – which he first called *shop management* and then *process management* – theorizing a vision of work that became known as **Taylorism**.

Confronted with the limits of artisanal, individualistic, and corporatist practices that were incompatible with the mass production enabled by new technologies, Taylor proposed dividing work both horizontally, with an optimal division of labour into specialized jobs, and vertically, with the separation between the design of methods and the distribution of tasks by engineers and managers, and their execution by operators.

The latter would no longer have to think but simply execute according to the instructions defined by the former following an analysis of production techniques (tools, work rates, postures, gestures, etc.) that formalize *"the one best way"* of carrying out the task.

Taylor did not particularly advocate assembly-line work, which already existed, but that workers be trained in the most efficient working method possible, using the most suitable tools and executing the optimum movements.

It was in 1913 that **Henry Ford** (after recalling a visit to the Chicago slaughterhouses as a teenager, according to his memoirs) introduced the horizontal division proposed by Taylor in the assembly of his Ford T, with each worker carrying out a single task as defined by the engineers, with the success that we know. Chassis assembly time fell from 728 to 93 minutes, and the selling price was reduced to 825 dollars compared with an average of almost 2,000 dollars for cars of the time.

The research on, and design of, tools adapted to the task, together with the optimization of movements, sequences, and work rates, can reduce the physical and cognitive constraints associated with the work, but they can also increase them if prevention is not integrated into the core of engineering work.

In any case, if it is carried out in a radical way, the limits of such an approach with regard to occupational health and safety appear on three levels:

- The number of jobs where the value added by the worker is confined to purely manual work is becoming increasingly limited.
- Even in this type of job, the discretion enjoyed by the worker, which is generally necessary for the task to be carried out properly (in terms of productivity and quality as well as comfort and safety), is systematically underestimated.
- What is more, this dehumanized form of organization, which denies all autonomy and reduces individual contribution to mere execution, seriously undermines motivation.

This **workers' discretion** means that, even when the task is simple, they cannot be treated like machines. Managing the workforce following the principles of D. McGregor's "Theory X" (McGregor, 1960) often ends in failure. On paper, supervision is always constant and effective; actions and their consequences can be traced back to the worker who carried them out, and the latter can be given appropriate feedback (reward or punishment in behaviourist parlance). The reality is often different.

If, on top of that, we explain to these workers that they are responsible for their own safety, when it is management who defines the means and methods of work, as well as the safety rules, their supervision, and the disciplinary procedures that go with them, this can only lead to frustration.

Key Learning Points for the Safety Leader

In very stable and/or highly regulated systems, errors, deviations, and workers' unsafe behaviour could, in theory, be avoided by technological and organizational solutions, based on perfectly defined methods and rules and betting on *rigour* in their application and supervision, according to the principle of **rule-based safety**.

Even if experience demonstrates the "sufficient" economic efficiency of this approach in some specific situations, it also shows its limitations, both in terms of effectiveness and sustainability – even for activities that are not very complicated – when it is not supplemented by the following safety vision.

MANAGING DYNAMIC SYSTEMS

Perrow pointed to the inevitability of major accidents in highly complex and tightly coupled systems, especially in transformation industries handling processes with high catastrophic potential (Perrow, 1984–1999). He theorized that problems can be successfully engineered out of simple and even of "complicated" systems, as long as the interactions of their components and subsystems are clearly defined and regular. But he insisted on the fact that complexity calls for uncoupling, leaving space for redundancies, buffers and slack.

As a matter of fact, even seemingly unsophisticated systems conceal unregulated activities and unsuspected interrelationships that call on the workers' judgement, which is essential when they are confronted with the job's "**reality**", defined by Dejours as all the difficulties against which will crash the protocols and instructions when executing a task (Dejours, 1995–2022). As soon as the processes and their environment are not under total control – and they almost never can be – the resulting variability will require adjustments on the part of the players, giving rise to behaviour that is difficult to predict and therefore almost impossible to regulate a priori.

As this variability is inherent to the task, its environment, and/or the resources involved (including human resources), it is unrealistic in practice to try to design and deploy control solutions that are solely technological or organizational in nature. We therefore have to rely on human intelligence, which, despite the limitations we are going to examine – and pending the possible emergence of "more proficient" artificial intelligence – is the only one capable of adjusting reactions to the emergence of volatile, uncertain, complex, and ambiguous phenomena (the famous **VUCA**).

This is why, in practically all work situations, the worker has a certain slack on the execution of established techniques and prior management planning, in particular regarding almost everything outside or beyond the "process".

The behaviour of "knowledge workers" and other technologists (more on these later) needs to be managed more in line with the principles of McGregor's "Theory Y" (McGregor, 1960), based on intrinsic motivation and trust, appealing to dialogue, participation, and empowerment mechanisms. These are the best ways to develop health, safety, and prevention as *values* on their own, as we will see later.

FOCUS ON... THEORY X AND THEORY Y

MIT management professor Douglas McGregor proposed in the 1960s two theories to explain workers' behaviours and motivation. Each one of them is grounded on a series of beliefs on the nature of humans' relationship with regard to work.

According to **Theory X**, human beings:

i. Do not like working, avoid it as much as they can, and must be obliged and constantly supervised to do it.
ii. Do not truly deploy their intelligence but to avoid the constraints imposed upon them.
iii. Do not like holding responsibilities, and prefer routine and being led.

However, according to **Theory Y**, human beings:

i. Are naturally inclined to deploy their industriousness and make efforts, are able to learn, and need to work to develop themselves.
ii. Are motivated by the desire to fulfil their potential through work if they are associated with the organization's objectives.
iii. Are susceptible of accepting (and even seeking) responsibility and prefer autonomy.

DID YOU KNOW? THE LIMITS OF DELIBERATION WHEN FACING COMPLEXITY

Logically analysing a complex situation means not only having access to a multitude of data but also having the cognitive capacity to manipulate them mentally by calling on our attention and our working memory, the limits of which we have seen. Then we also need references against which we can value the scenarios that we imagine and the efforts that their implementation will require.

The ability of human beings to carry out these processes has been the subject of a great deal of research. Wilson and Schooler showed that people can actually do worse in tasks involving decision-making when required to analyse the reasons of their preferences or to evaluate in detail the different aspects of the possible alternatives (Wilson and Schooler, 1991). Particularly surprising were the results of the experiments conducted by Ap Dijksterhuis, who asked a series of subjects to imagine that they were to buy a car.

Half the participants were given simple presentations of 4 cars (with only 4 features described) and the other half more detailed specifications (with 12 features described). In each group, the first sub-group was asked to concentrate on the analysis for 4 minutes to make their choice, while members of the second sub-group were given distractions to entertain them during the 4 minutes preceding the moment of choice.

Among those who received the simplified descriptions, it was those who concentrated on the analysis who made the best choices. But among those who received the detailed specifications, it was those who had been distracted from the analysis who chose the best model! (Dijksterhuis, 2007)

On the basis of these and similar experiments, some have postulated the superior ability of our subconscious mind to deal with complex problems. The reason may lie in the fact that it mobilizes neural structures that are far more powerful than those supporting our attention and working memory. The problem is how to ensure that these processes, which by definition are beyond our conscious control, are systematically oriented towards safety...

Key Learning Points for the Safety Leader

In most work situations, the concept of **rule-based safety**, characterized by the obligation to comply with rules and instructions, must be supplemented by that of **managed safety**, which requires a good understanding of the human factor.

It is on the basis of such knowledge that we can develop a *culture of safety* – using the approach and the tools that we will see. Once such a safety culture is established, we will be able to appeal to the *integrity* of every worker, which will guide their conscious reflection and deliberate choices, as well as their "intuitions" and other spontaneous choices, through mechanisms that we will examine below.

LOGIC

"It's always easy to be logical. It's almost impossible to be logical all the way through".

Albert Camus, Le mythe de Sisyphe (1942)

"All our reasoning is reduced to giving in to sentiment".

Blaise Pascal, Pensées (1670)

The employer therefore needs the workers' intelligence to carry out their work and expects them also to use their intelligence to avoid injury and, more generally, to prevent accidents.

The Cartesian vision of our rationality would have us believe that all the data accessible to our brain from the processes described in the previous chapter should be processed in a **logical manner**. However, the very concept of logic can be approached from different angles. From Aristotelian syllogism to the various forms of non-binary logic, from algorithmic calculation to propositional language, the notion of logic remains subjective, evolving, and cultural (Blanco-Munoz, 2019).

FOCUS ON... NEOCLASSICAL ECONOMICS' LOGIC

Since the end of the 19th century, the dominant neoclassical economic theory has been based on the notions of marginal utility and market equilibrium, which are in turn based on a conception of human beings as *homo economicus* (Veblen, 1898; Clerc and Piriou, 2011).

Stripped of feelings and emotions, in line with the scientific rigour of the time (the aim was to give economics the status of a science on a par with physics), it was assumed that an individual's behaviour is fully rational and guided by the following three principles:

- **Maximization:** Individuals constantly seek to maximize their satisfaction (utility) and minimize their costs, in a perpetual quest to optimize their limited resources.
- **Consistency:** Their choices are guided by preferences for different factors that can bring them satisfaction, preferences that are stable and that they can prioritize (transitivity).
- **Independence**: Their preferences and therefore their choices are independent of any social or historical influence (sovereignty).

It should be noted that the above postulates implicitly assume that *homo economicus*:

- Are supposed to have access to all the information necessary to analyse objectively the risks and opportunities associated with the consequences of the various behaviours they may adopt,
- Possess the cognitive skills and methods to carry out this analysis,
- Systematically take the time to perform this analysis,
- And thus, unless uneducated and/or stupid, are rarely wrong.

This conception of the human being has no positive application in the field of prevention; on the contrary, it has traditionally stood in the way of the development of a culture of prevention as presented in the following chapters.

In any case, our ability (and willingness) to use logic is not perfect or even systematic. Contrary to the neoclassical economic vision, we are not omniscient and enlightened maximizers, but rather we are content to satisfy our needs and desires as best we can by devoting what we consider to be the minimum cognitive resources necessary (attention, concentration, reflection, time, etc.) as famously put by Herbert Simon in the 1950s (Simon, 1947, 1956).

Key Learning Points for the Safety Leader

In a world populated by people guided by perfect instrumental rationality, there would be no accidents in principle. Everything that should have been logically anticipated would have been, and day-to-day operations and management would be in the hands of individuals capable of making the best choices to protect their health and safety.

Any accident would come as a surprise and would necessarily imply gross negligence on the part of an individual who should and could have acted "correctly" but did not. This approach may seem gratifying in the short term for the organization and its management, as it relieves them of responsibility, but it provides no tool other than the application of disciplinary measures aimed at the person "at fault", and therefore no real method for implementing preventive or even effectively corrective actions.

Managers and institutions who rely on this vision to build their prevention strategy are systematically frustrated by their health and safety performance and fail to find ways of improving it. Unfortunately, it still applies today and is reflected in the search for those responsible and at fault in the event of an accident.

HEURISTICS

The first limitation of our deliberative processes is our natural tendency to short-circuit them. Even if we are capable of logical reflection, our brains operate most of the time on a day-to-day basis by **recognition**. Skills that are often referred to as expertise, judgement, or intuition are in fact based on our ability to subconsciously recognize signals in a given situation: elements and/or patterns that are already known and which, by *association* with our memory, bring up information about how to act in situations in which such signals are present (Simon, 1972).

These fundamental human qualities have been studied in detail since Simon revealed their nature. Gary Klein presented them elegantly as part of what he called *naturalistic decision making.* His field research demonstrated that, most of the time, our decisions do not come from the comparative evaluation of the utility and probability of the different options we can choose from. According to his Recognition-Primed Decision Model, if when experiencing a situation we do perceive it as typical (either prototypical or analogue to a prototypical one), the mere recognition of a familiar pattern instantly produces four by-products:

- *Relevant cues* consistent with that pattern become evident; we tend to focus our attention on those and to ignore others to avoid the paralyzing effects of ambiguity and information overload.
- *Expectancies* about what may happen next appear in our mind.
- *Plausible goals* related to what we usually do in such instances are established, that we use to set priorities.
- *Typical actions* or ways of responding to achieve those goals occur to us.

Typical course of action is then implemented in the case of a simple match. We do not have the impression of actually deciding anything: this process unfolds subconsciously and we just act. And this applies to all levels, individually and collectively. The research of Paul Nutt from Ohio State University on organizational decision-making, cited by Steven Johnson (Johnson, 2018) revealed that only 29% of the organizational decisions studied contemplated more than one alternative.

If a simple, single match is not immediately available in our minds, possible choices are then searched for, identified and considered. But in general, we do not proceed to a comparative evaluation of all of them simultaneously, it would be too costly in terms of time and cognitive resources. We rather adopt a *singular evaluation approach*: each option is evaluated independently on its own merits, in a serial manner following Simon's *satisficing* principle. If the first possible action seems to fall short of achieving the typical goals, we move on to considering a second one, and so on.

We can relate this to the **practical reason** (*phronêsis*) that Aristotle proposed as one of the five human intellectual virtues, and that differs from "understanding" (*synesis*) and "good sense" (*gnome*) as these two only "pass judgement", whereas the former "issues commands".

This mode of functioning by means of **heuristics** or *mental shortcuts*, which Daniel Kahneman called "System 1" (Kahneman, 2011), saves us the mental effort required for reflection and allows us to act faster. It is valid and helps us to be more efficient in contexts where:

- The situations are ambiguous and uncertain, goals are unclear, procedures are poorly defined and information is incomplete or unreliable (which we could associate to fairly loosely coupled and relatively complicated or outright complex systems in Perrowian jargon).
- An approximate solution to the problem or a rough estimate of the result is acceptable, either because of urgency, time-constraints or low stakes (heuristic relevance).
- The situations are dynamic, but the overall long-term conditions of the system and its environment are stable or change slowly.
- The individuals can acquire experience by sampling a significant number of situations and identifying common elements and patterns.
- The individuals receive immediate and precise feedback confirming or denying the relationship between the signals, the adopted behaviour, and the results obtained with regard to those expected.

This feedback validates or invalidates the appropriateness of the behaviour adopted, and its repetition results in inductive learning, which will be chronically accessible at a subconscious level through the reinforcement or long-term potentiation of the synaptic connections involved, as described by the neuropsychologist Donald Hebb (Hebb, 1949).

So, this quality of ours is precious when we need to decide quickly to act fast, the same way as our sensorimotor skills, our reflexes, and our procedural automatism that we studied before. But very much like those, our intuitions and expertise are never infallible and can lay traps for us. The conditions listed above imply that this mode of reasoning may be particularly valuable in VUCA contexts when time is of the essence. But it may be unsuited in high-risk activities requiring a high level of precision or certainty. However, once we have acquired, through experience, the conviction of the validity of our mental shortcuts, we tend to apply them even in dangerous situations.

We call this excessive tendency to follow familiar mental paths the "law of the instrument", *Einstellung* effect, or Maslow's hammer, although it was Abraham Kaplan who first wrote, "give a small boy a hammer, and he will find that everything he encounters needs pounding" (Kaplan, 1964). He illustrated with this expression our predisposition to apply solutions (or mental tools) that we know well without considering alternatives, as if we were wearing blinkers.

FOCUS ON... EMOTIONAL INTELLIGENCE

Our emotions, whose role as triggers for action will be presented later, are also decisive in our higher order cognitive processes. Among other things, a particular region of our brain called the insula constantly scans the signals felt by our body in a process known as **interoception**. A multitude of circuits then act at a subconscious level to analyse these sensations and produce affects (pleasant/unpleasant, calm/agitated, etc.), the precursors of emotions.

Processing these sensations and affects involves a system of valorization that assigns them a positive or negative value, which precedes conscious thought (Damasio, 1994; Teboul and Damier, 2022). Recognizing a familiar pattern in a situation would trigger an internal satisfaction feeling, which would preclude our motivation to spend additional time and mental energy in thorough analysis or seeking further information.

We will come back at length to this notion of valorization and its crucial role in our attitudes and behaviour.

DID YOU KNOW? A SECOND BRAIN IN OUR GUTS

Have you ever experienced what are known as "gut feelings", such as a feeling of anxiety fluttering around in your stomach whose origin you cannot precisely identify?

One of the most widely discussed examples is that of the Cleveland firefighter reported by Gary Klein (Klein, 1998) who, on the basis of a gut feeling, suddenly ordered his team to leave the burning house they were intervening on, seconds before it collapsed. After this episode, this lieutenant believed for years that he possessed a "sixth sense" or "extrasensory perception", until Klein offered him a more rational explanation with his Recognition-Primed Decision Model following their interview.

Nevertheless, it may very well be in our guts that we "feel" that something is odd.

The enteric nervous system (ENS) is made up of two layers that line our digestive tract from the oesophagus to the rectum and contain more than 100 million neurons. Recent discoveries about what is now known as the *brain–gut axis* show that the ENS does more than just regulate the secretions and motor functions associated with digestion: connected to the central nervous system, it plays an important role in regulating our mood and, via our emotions, certain cognitive functions such as thinking and memory.

This research, which is still in its infancy, is beginning to shed light on how subconscious affects and thoughts can trigger secretions in the digestive system via the ENS, which would in turn act as a "sounding board" for what are known as "intuitions" (Gershon, 1999).

Key Learning Points for the Safety Leader

Rapid mental modes of operation based on **association** and **valorization** of our sensations are generally justifiable from a cost/benefit point of view, but they are far from infallible: signals may be missed, misperceived, or misunderstood, or other parameters that are not part of the subconscious mental model developed by the individual (which we will present further below) may come into play in a specific instance.

For better or for worse, in the absence of any particular motivation (a mechanism we will come back to later), we naturally tend to follow these mental shortcuts because they save us the effort of reasoning, including in hazardous situations.

COGNITIVE BIASES

It is again Daniel Kahneman, with Amos Tversky, who initiated the research on **cognitive biases** (Kahneman and Tversky, 1974): they experimentally demonstrated systematic distortions and deviations from rational behaviour as we might expect it.

Some of these biases relate to the sensory and neurological characteristics of the human being and have already been mentioned (*sensory-motor, attentional,* and *memory biases*). These biases also come into play when the individuals feel the need to analyse the situation deliberately, using their "System 2", which is also affected by *reasoning* and *judgement* biases.

All these families of biases are adaptive solutions to three problems:

  i. Too much information regarding our cognitive capabilities,
 ii. Lack of meaning regarding our paradigms, and
iii. Need to act quickly.

More than a hundred cognitive biases have been documented. Below are presented briefly some of the most significant ones with regard to occupational health and safety.

Loss Aversion

We do not like the feeling of losing something, including our time and energy.

We may be tempted to neglect certain precautions – which represent an immediate and definite effort – in the face of damage that seems unlikely, or simply remote in time.

FOCUS ON... PROSPECT THEORY

In the 1970s, Amos Tversky and Daniel Kahneman demonstrated our cognitive biases and heuristics, or mental shortcuts, which affect our decision-making when confronted with uncertainty. They formulated their famous **prospect theory** (Kahneman and Tversky, 1979) according to which, when we consider choices whose consequences are uncertain, it is not the possible final states (the total utility expected at the end of each scenario) that are evaluated by the individual but rather the prospect of gain or loss (the utility delta).

Experimentally, they demonstrated our *loss aversion*, which is notably characterized by:

i. The fact that a potential loss is more unpleasant to the individual than the equivalent potential gain is pleasant: for example, finding a 10-dollar bill brings us less happiness than losing 10 dollars brings us grief.

ii. The diminishing psychological value of gains and losses: for example having to spend 10 minutes on safety checks if we normally spend 5 minutes on them (i.e. 5 minutes "lost") upsets us more than having to spend 35 minutes if we normally spend 30 minutes on them (even if in absolute terms the additional amount of time is the same).

iii. *Risk appetite* when it comes to avoiding or reducing losses: for example, an investor or bettor will take more risks to make up for losses than she would to increase her gains.

The diminishing utility curve, well known to economists, is still valid for gains, but it is not symmetrical on the side of losses, as the slope is much steeper on this latter side (Figure 44).

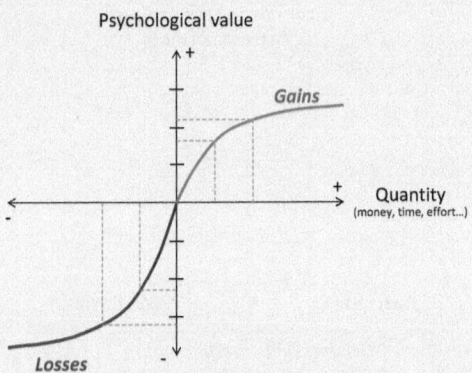

FIGURE 44 Illustration of Kahneman and Tversky's prospect theory (1983).

Therefore, and against conventional wisdom, this bias is not in favour of prevention. On the contrary, it explains why we accept the probability of damage, even severe injury, which seems remote and uncertain to us, rather than the constraint of prevention because it seems a loss to us (of time, comfort, energy, etc.) which is immediate and certain.

Status Quo Bias

The risks associated with change or novelty are considered more important than the benefits they could bring. Linked to the loss aversion mentioned above, this bias is often an obstacle to the implementation of preventive and corrective actions, and more generally to change initiatives, even if they bring improvements in terms of safety.

Dunning–Kruger Effect

The tendency to overestimate our qualities and skills is a virtually universal bias. Most people tend to see themselves as better than they really are in a wide range of parameters like beauty, strength, resilience, and professional competence. It is very difficult to objectively assess oneself in domains where there are no conventional quantitative metrics to compare one individual to another.

TEST NO. 7: Are you a good driver?

In the light of what we have seen in the previous sections, think about your driving skills. Picture yourself behind the wheel and consider how well you perceive your environment and eventual risks, your ability to anticipate, your sensorimotor intelligence and reflexes, and your capacity to remain focused and alert...

Now, answer honestly this simple question: **do you think you are among the top 50% of drivers, or that you belong to the lower half?**

Between 80% and 93% of drivers surveyed believe they are in the top 50% of drivers (Figure 45). There are therefore many who overestimate their abilities, and some by a considerable margin (Goszczynska and Roslan, 1989).

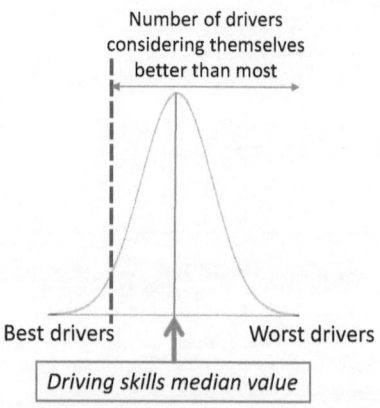

FIGURE 45   Driving skills self-estimate.

However, it appears that it is the least competent individuals in a particular field who tend to relatively overestimate their competence the most. The beginnings of mastery, which are enough to feel at ease in "normal" circumstances but not yet sufficiently developed to be aware of the difficulties associated with "abnormal" or simply unknown circumstances, can lead to **overconfidence** (Kruger and Dunning, 1999). Although the original research of Kruger and Dunning has been recently questioned, the validity of their findings remains widely accepted.

Expert Bias and Curse of Knowledge

If beginners, as soon as they have gained some confidence, can lure themselves into believing that they are in full control, so can "experts". At the opposite side of the learning curve, experts eventually notice that, after many years in their trade, the progression in their skills tends to slow down: it approaches an asymptote that will never reach total infallibility, but they may get the impression that there is nothing left for them to learn.

Their deep knowledge can make them fall prey to the heuristics discussed above, as well as reduce their openness to alternative perspectives or new information. They can stop questioning themselves, and they lose motivation to go through preventive routines, double-checking, and seeking second opinions.

Expert bias appears to be particularly prevalent in medicine, where doctors are highly respected and deal with complex situations and ambiguous data, but can also be observed in many other professions where seniority and experience are regarded as a proxy for wisdom.

Another potentially negative side effect of expertise is the **curse of knowledge** (aka the curse of expertise). Professionals with specialized knowledge assume that other, less proficient individuals around them share in that knowledge. Managers, supervisors, and senior staff find it difficult to put themselves in the position of more junior, inexperienced, and unskilled team members. Therefore, they may provide summary explanations of how the task must be performed and/or the hazards that it implies, as well as insufficient resources (tools, material, time, support…) to complete it (Camerer et al., 1989).

Availability Heuristic

Subconsciously, we estimate the frequency of occurrence of an event (and consequently the probability of future occurrence) on the basis of the ease with which we can access it in our memory. However, we have already seen the limits of our declarative memory: this mechanism can therefore lead us to underestimate risks.

We may therefore consider a risk to be very low simply because at the time we cannot summon any "striking" memory, such as the insistent and repeated reminders and injunctions from management or, way more memorable, an accident in similar circumstances involving ourselves or someone close to us.

TEST NO. 8: What are you most afraid of?

Look at the list of accidental events below: they are arranged in pairs (Figure 46).

Accidental poisoning **Vs.** Accidental drowning

Lightning **Vs.** Airplane accident

Hornets, wasps or bees **Vs.** Choking on food

Earth movements **Vs.** Fireworks accident

Smoke, fire or flames **Vs.** Dog attack

FIGURE 46   Assessing risks, 1 on 2.

For each pair, try to estimate which of the two events would be most likely to kill you or someone close to you one day:

Once you are done, you can proceed to the next paragraph.

Here are the actual statistical probabilities in the United States, in ratio to the total number of accidental deaths per year (boxes), and in relative proportion to the other event in the same pair (fraction) (Figure 47):

Accidental poisoning **Vs.** Accidental drowning

| 1 in 139 | 8 / 1 | 1 in 1,073 |

Lightning **Vs.** Airplane accident

| 1 in 81,701 | 4 / 1 | 1 in 354,319 |

Hornets, wasps or bees **Vs.** Choking on food

| 1 in 62,950 | 6 / 1 | 1 in 370,035 |

Earth movements **Vs.** Fireworks accident

| 1 in 153,597 | 7 / 1 | 1 in 1,000,000 |

Smoke, fire or flames **Vs.** Dog attack

| 1 in 1,235 | 97 / 1 | 1 in 119,998 |

FIGURE 47   Assessing risks, 2 on 2.

Source: USA National Safety Council, statistics on accidental deaths in the US in 2006, except for air crashes and suffocation, from 2001.

As we have seen, our long-term memory does not function as a database giving us access to the quantitative information that would enable us to make statistical calculations and thus objectively assess risks. In everyday life, all we have is the "feeling" or "intuition" that something is more or less likely.

But this feeling is based on the more or less strong imprint that these events have left on our memories. And this imprint depends not only on the frequency with which we were exposed (directly or indirectly) to these scenarios but also above all on their emotional load. As we saw in the section on memory, the consolidation by the hippocampus of memories of emotionally loaded events will be reinforced by the amygdala.

DID YOU KNOW? ON SHARKS AND MOSQUITOES

Any untimely death is tragic, but some attract more media attention than others. Mosquitoes kill around 800,000 people a year worldwide, mainly in developing countries, carrying malaria, yellow and West Nile fever, dengue fever, Zika, and chikungunya. However, few people in the Western world pay much attention to these diseases.

We are more impressed by the news and stories about shark attacks (Steven Spielberg's 1975 film "Jaws" also contributed to this apprehension), even though they only cause around ten victims a year.

Our subjective assessment of the risk linked to mosquitoes and sharks is biased by this relative emotion, which means that these shark attacks are more available in our memory.

Base Rate Neglect

We rely on the availability heuristics described above and overestimate the validity of our predictions because, in general, we ignore the statistical probability of the phenomenon in question. But even when we have an idea of the odds, we often consider them irrelevant to our particular case. Oftentimes this is due to unrealistic optimism or plain wishful thinking, but sometimes we do have data and instances that actually support our views. The problem is that we can bank on such data, issued in general from our own experience without weighting the size of the sample that our existence has provided us with.

It is easy to be confident in the reliability of a method, a system, or a component, or conversely to underestimate a risk, based on a number of occurrences that is not necessarily statistically significant.

TEST NO. 9: Who needs protective gloves?

Imagine you are a health and safety officer in a factory and you are analysing the accident figures for the past year.

Out of a population carrying out an activity involving the handling of sharp parts, you have had 100 cases of cuts to fingers. In these 100 accident reports, you note that in 70 cases the worker was wearing safety gloves, as instructed (Figure 48).

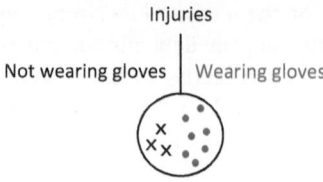

FIGURE 48 Injuries and gloves, 1 on 2.

Since more workers are injured with gloves than without them, some workers see this as a reason to drop the obligation to wear gloves, which they see as a constraint (loss of dexterity, discomfort), and your management is questioning the necessity of continuing to supply these gloves, which are expensive.

WHAT DO YOU THINK?

Without going into statistical calculations, Figure 49 illustrates that it is much less likely to be injured when working with gloves than without them, and therefore the benefits of wearing gloves.

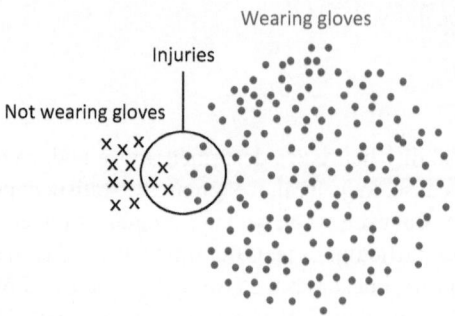

FIGURE 49 Injuries and gloves, 2 on 2.

Psychophysics of Chances

Our subjective assessment of the increase (or decrease) in the level of risk does not linearly correspond to the mathematical probability of occurrence of the adverse event.

Going from absolute safety (0% probability of injury) to a 5% probability of injury upsets us more and makes us react more strongly than going from a 30% to a 35% probability of injury (Figure 50).

Confirmation Bias

We subconsciously sort information, selecting that which confirms or supports our *beliefs* and *values* (concepts that we will develop later). This **confirmation bias**, a term coined by cognitive psychologist Peter Wason (Wason, 1960) (aka myside bias or congeniality bias), is particularly prevalent in VUCA contexts where the validity

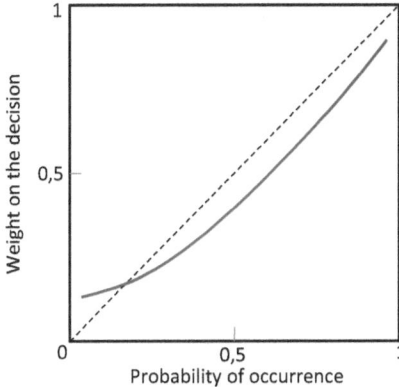

FIGURE 50   Psychophysics of chances. (Adapted from Kahneman and Tversky (1984))

of available information is subject to interpretation. And today, as the Internet makes all sorts of opinions widely available, we tend to cherry-pick the ones that are in line with our own.

Confirmation bias concurs to phenomena such as the *irrational primacy effect* (relying more strongly on information that we come up with early), *belief perseverance* (reluctancy to changing our minds even when presented with contradictory information), and *illusory correlation* (finding associations between unrelated events or situations).

If we are convinced of the safety of a situation, this opinion will be reinforced by indications pointing in that direction, whereas we will tend to ignore or underestimate the relevance of information pointing in the opposite direction.

Groupthink, Conformism, and Bandwagon Effect

Manifestations of our tribal reflexes, groupthink, conformism, and bandwagon effect are three related biases that push us to adopt certain behaviours because others around us do so. They derive from deeply ingrained mental structures that developed to facilitate social life and individual survival by avoiding exclusion.

Groupthink is mainly driven by the desire of the group members to maintain harmony and avoid disagreement. However, seeking cohesiveness at any cost may result in dysfunctional decision-making processes; for instance, if everyone aligns with the first one giving an opinion or to the advice of the more respected person in the room (whether because of hierarchy, charisma, or seniority) resulting in flawed outcomes (Turner and Pratkanis, 1998).

Whereas **conformism** makes us follow behaviours that we observe others do, and its mechanisms might be connected to the mirror neurons that we mentioned before, the **bandwagon effect** relates to the fact that we are pulled by popular ideas and public opinion. We tend to overestimate the relative value of the members of our group and their abilities and align ourselves with the dominant view in our group.

These biases can inhibit the individual's critical assessment of risks and their motivation to express fears, challenge methods, or point out dangers that would be generally accepted by other team members.

FOCUS ON... SYSTEM 1 AND SYSTEM 2

Loss aversion and the other biases and heuristics described by Tversky and Kahneman and largely studied and commented since (Kahneman and Tversky, 1984; Kahneman, 2011; Thaler, 2016) shed a new light on judgement and decision-making. They review and correct, from a *descriptive* perspective, the *normative* view of behaviour given by neoclassical economics.

Observation and analysis of the various apparently irrational (or anomalous) yet systematic behaviours and choices led them to theorize the existence of two mental operating modes or "systems" for decision-making. **System 1**, instinctive and emotional, operates by association and prioritizes the process's speed. **System 2**, reflective and logical, operates more by deduction and gives priority to the rational quality of the analysis.

Although we have the intellectual capacity to switch to System 2 as soon as we feel the need to do so, it appears that most of our behaviour is guided by System 1. Natural selection has favoured the development of a prefrontal cortex enabling us to deliberate but, as we have seen, has also preserved (and developed) neuronal structures in us capable of automating our reactions as far as possible, thus prioritizing speed of reaction over reducing the error rate.

As a species evolving in a context of threats and opportunities, it is more effective to be right 90% of the time, quickly, than to be right 99% of the time, but too late.

Key Learning Points for the Safety Leader

The aversion to wasting time and energy taking precautions against damage that is never certain is a very powerful bias. Even if we are aware of the danger, if we think that all we have to do is "pay attention" and "be careful" to avoid an accident, what is the point of wasting time and/or resources by taking precautions that would be redundant and therefore pointless in our eyes?

The combination of this aversion with the underestimation of the probabilities of "having a glitch" and the overestimation of our abilities discussed above explains a good number of behaviours that may appear as unreasonable to the preventive observer and which can hardly be combated by simply appealing to deliberation. Other approaches exist and will be presented later.

RATIONALITY

Since we are generally unaware of the effects that the above shortcuts and biases have on our cognitive processes (and are not free from their influence even when we are aware of them), we make decisions that seem reasonable to us on the basis of the information we have at the time, according to what Herbert Simon called the principle of **bounded rationality** (Simon, 1972) (Figure 51).

FIGURE 51 Local rationality (drawing by author).

This principle, closely related to that of **local rationality** (Eurocontrol, 2014), makes human beings adopt behaviours that are, in their view, fully rational at a given moment, taking into account:

- The *focus of their attention* (as previously seen),
- Their *knowledge* (addressed in the following section), and
- Their *objectives*, which exist in a particular context and, like this one, are dynamic. They are often different from the objectives "stated" by the individual and by the group, and may therefore not correspond to the work as prescribed.

With regard to this last point on objectives, it is assumed that individuals wish to carry out their work in a correct and efficient manner, thus also taking into account the objectives of the organization.

Although this can be accepted as being generally true, in real life workers will try to reconcile personal needs and motivation factors (comfort, recognition, fulfilment... and safety) (Maslow, 1943) with very diverse, evolving, and sometimes conflicting organizational rules and expectations (productivity, efficiency, quality... and safety). Each individual is constantly making trade-offs and adjustments to meet each of these objectives to the best of their abilities, a perpetual balancing exercise that Erik Hollnagel calls ETTO (*efficiency–thoroughness trade-offs*) (Hollnagel, 2009).

FOCUS ON... THREE TYPES OF RATIONALITY

When we talk about rationality, we often think only of the logical dimension of utility, according to the principles of *homo economicus* presented above, for whom the rationality of behaviour is based solely on a comparison, supposedly objective, between the means at their disposal and the objectives they are pursuing.

German sociologist Jürgen Habermas expanded this concept by postulating the existence of three types of action, which call on three types of rationality:

- Instrumental action, directed towards a material objective, subject to **instrumental rationality**, whose validation criteria are truth and efficiency.
- Moral-practical action, oriented towards relationships with others, subject to **axiological rationality**, whose validation criteria are goodness, justice, and equity.
- Expressive action, oriented towards the staging necessary for understanding by others, subject to **dramaturgical rationality**, whose validation criteria are authenticity and expressive coherence.

Each of these three types of rationality is subjective and can only be validated by *intersubjective consensus*. All three are employed at work to varying degrees in different situations, and even purely technical gestures, aimed primarily at utility and efficiency, incorporate an axiological rationality based on values, which will be discussed below.

Thus, when we consider these three types of rationality, what seemed irrational because it was too dangerous may appear to correspond to certain values of the individual and/or the group and therefore rational at this level (Habermas, 1987).

Classic decision theory, which has interestingly also been called "theory of rational choice", only considers the first one of the rationalities listed above. It does not care much about social and cultural rationality, a concept dear to Perrow that we will revert to in the following chapters. When facing uncertainty, the "rational agents" would focus on expected utility and probability to evaluate options and make optimal decisions. They would identify all relevant variables, then build and probe all possible scenarios (or at least the most significant ones). We have seen that this is extremely rare. Even when motivated to deploy such efforts we cannot escape the influence of the heuristics and biases described.

What Klein proposes is that, besides the sequential mental simulation of the unfolding of different courses of action, the best that we usually do is to try and clarify the anomalies that we may detect in the pattern that we have recognized. Far from using sophisticated analytical methods, we tend to rely on *story building* to justify those anomalies. We do not let go of the idea of matching the situation we thought we recognized until such story becomes exceedingly untenable.

As we will develop, we have better chances to identify the flaws in our reasoning when exposing it to others.

DID YOU KNOW? THE ADVANTAGES OF DIVERSITY

Health and safety specialists, and in particular those with expertise in machine and process safety, are well aware of the importance of redundancy and the **diversity** of safety devices. If you have a tank and your HAZOP tells you that overfilling can cause problems, doubling the level probe by fitting a second float is not the best idea. If residues can clog up the circuit of the first float, it is likely that they will also clog up that of the second. It is better to use an ultrasonic sensor (for example).

Similarly, the diversity of rationales among workers is of major importance and is often underestimated. At all levels and in all functions (operators, managers, designers, etc.), integrating people from different backgrounds into your organization gives you a fresh perspective. Workers from other organizations, other sectors of activity, and more broadly, other cultures (as we will define them later) can bring with them not only "ready-to-use" good practices but also, and above all, a different way of approaching problems.

Key Learning Points for the Safety Leader

The safety leader must always bear in mind that any decision taken by an individual stems from their rationality, which is by definition local and limited, and that it will never be purely instrumental since it includes an ethical dimension, which we will study later.

Thus, the level of **discretionary power** that individuals can exercise at work when adapting to the task, its constraints, and their multiple objectives must be studied very carefully.

A certain degree of flexibility or *slack* will always be necessary for the worker to cope with the variability of the operating conditions, the work environment, or the tasks themselves. We will therefore have to assume a certain variability in their behaviour. As we saw above, this room for manoeuvre, which is usually informal and poorly defined or not at all, exists even in the most regulated activities.

KNOWLEDGE

The knowledge available to the individuals and the way in which they assess its relevance and completeness in the context of the choice to be made will be decisive for their safety. Whether explicit and accessible for methodological evaluation, or tacit and only "felt" in the form on an intuition that one cannot explain, its importance is capital for ensuring safe operations.

FOCUS ON... MANUAL WORKERS VS KNOWLEDGE WORKERS

The traditional division between manual workers and knowledge workers was a handy way of codifying the skills and competences needed to do a job, as well as the types of mistakes that could be made. The former mainly used their System 1, while the latter were supposed to make greater use of their System 2, as described above.

However, this dichotomy is no longer valid, at least in industrialized countries where labour costs and new technologies mean that any task that could potentially be automated has been or is destined to be automated. Most workers today occupy intermediate positions between the "traditional" manual worker and the engineer or manager. Peter Drucker called such workers **technologists** (Drucker, 1974): even if a large part of their time is devoted to reproducing tasks and gestures, they also take initiative and must have the skills and knowledge to do so.

Knowledge management thus emerges as a new discipline, which has been associated – and even integrated by some – with organizational learning (Senge, 2006) and complexity theory (McElroy, 2000). The following sections summarize the main difficulties in this area that could have consequences for safety.

Lack of Knowledge

The lack of knowledge required to perform a task is an obvious factor that can lead to errors and accidents. Even though Henry Ford once said that "any man can learn anything he will, but no man can teach except to those who want to learn", we may need to nuance the first part of his statement. The individuals' cognitive capabilities may be a limitation to their ability to effectively learn more or less complicated concepts and skills. Through a set of standardized tests, the intelligence quotient has traditionally been used to measure human intelligence. But even if it is not done in such a formal way, all human resources selection processes aim to determine the candidate's cognitive aptitude to meet the job requirements.

Once we have ascertained this preliminary point, we can agree with the insightful second part of Ford's quote: learning requires an effort and hence motivation (more on this later), and the cognitive biases described above can "lock in" the minds of certain individuals.

But in an organizational context, any lack of knowledge on the part of a worker must be analysed as a failure on the part of the organization to capitalize on and transmit this knowledge, a process that must go as far as ensuring that this knowledge is actually acquired and systematically used by each individual. Every organization must have the ambition to become a *learning organization*, as we will describe them later.

However, it is also important to note that, as Nonaka and Takeuchi put it: "Knowledge in the strict sense can only be created by individuals... The creation of organizational knowledge must be understood as the organizational process that amplifies the knowledge created by individuals and crystallizes it as part of the organization's knowledge network" (Nonaka and Takeuchi, 1995).

Two types of knowledge will emerge and must be capitalized on and developed (Polanyi, 1983):

- *Explicit knowledge* (related to declarative memory presented above) which can be communicated by language and documented. It is typically the subject of more or less formal training and qualifications within organizations when it is necessary for risk management by workers.
- *Tacit knowledge* (related to phronêsis, synesis, and gnome previously mentioned) is acquired through experience and is often intangible. Its importance is systematically underestimated. Because of this, and because it is difficult to codify, it is most often passed on informally between peers.

They cannot be dissociated: tacit knowledge is a prerequisite for explicit knowledge, because tacit knowledge provides the context of meaning from which the individual acquires explicit knowledge. And this implicit context emerges from a network of cultural conventions that we will look at later.

There is a tendency to overlook the importance of tacit knowledge, which is often essential, well beyond this notion of contextual information enabling interpretation: like genetic information in an organism, explicit information within an organization does not act without being activated by tacit knowledge (Keller, 2000).

DID YOU KNOW? COMPETENCE

It is from the integration of explicit and tacit knowledge that **competence** emerges, defined as the ability to put knowledge and know-how into practice to achieve the desired results (ISO/IEC).

Key Learning Points for the Safety Leader

Formal training programmes will never be able to cover all the knowledge, know-how, and skills needed to perform a task – let alone a job – perfectly.

Faced with this reality, many organizations set up mentoring programmes with varying degrees of supervision, ensuring that new recruits are accompanied by an experienced worker during the critical phase of building their skills.

These programmes also make it possible to showcase these experienced workers, which is a recognition factor that helps to motivate them (more on this later).

Illusion of Understanding

One of the consequences of the various biases presented above is that we think we understand the world in general – and the various phenomena we encounter on a daily basis in particular – much better than we actually do. Most of us think we have a good understanding of geopolitics and international relations or how a toilet flush works, but when we are asked for concrete details, we do not necessarily know them. In fact, we have a "sufficient" understanding depending on our role, which is generally that of spectator or, at best, basic user.

In a world of work where sophistication is growing exponentially at technological, technical, and organizational levels, workers are increasingly disconnected from the basic knowledge behind the techniques and artefacts they use. In theory, this should not be a problem in a Taylorist organization, where certain experts take care of the installation and maintenance of a machine that the operator simply has to operate. But the intuitive and approximate knowledge of the operator with regard to certain mechanisms, or of the first-level technician who tries to "cobble together" a solution thinking she understands the problem, can give rise to dangerous conditions or directly cause an accident.

This illusion of understanding is even more glaring when it comes to human performance. Everyone thinks they understand how people's minds work, what motivates them, and why they act the way they do. We hope that this book will help to dispel some of the most common misconceptions.

Managers and leaders' illusion of understanding human factors at work is at the root of instructions, communication programmes, management systems, compensation and discipline schemes, policies, and more generally, paradigms (discussed below) that are not only ineffective but can be counterproductive and even dangerous.

Last but not least, this phenomenon can simply interfere with communication. Symmetrically to the *curse of knowledge* affecting the emitter as presented above, the receiver of a message may think that they have picked up and interpreted all the information they need and that they have perfectly understood what they have to do, whereas they would only have too general a view to effectively resolve certain details of the task.

Workers may consider that they have understood a written or verbal instruction but find themselves in difficulty when carrying it out and trigger or otherwise contribute to an accident.

Key Learning Points for the Safety Leader

As operators, managers, and executives, we all believe that we have a much better understanding of the subjects we work on than we actually do. This approximate understanding can lead to technical, tactical, and strategic errors (to which we will return below).

Inadequate Paradigms

Another phenomenon emerges from the integration of our knowledge: as it weaves its way through our psyche, it gives rise to our paradigms. We must consider that (Meadows, 2008):

- Everything we think we know about the world is a model: every word in every language, every map and every statistic, every equation and every piece of software. Just like the perception that we have of the world, we build our **mental models** as much on our own experience as on the basis of partial models acquired within the *moral circles* to which we belong and which we will describe later.
- Our models are all the more useful if they are congruent with the world in general and with the systems with which we interact in particular.
- But these models are only representations, not reality. The very need for these models stems from our inability to perceive and understand the world as a whole. As we have seen, we can only manage a limited number of variables. Since they are incomplete, these models are imperfect and can mislead us.

These errors are all the more surprising given that the actions resulting from conclusions based on our models have proved successful in the past.

Errors and accidents can therefore arise just as much from drawing logical conclusions from an insufficiently precise model (*induction problem*) as from a flaw in the logical process itself (*deduction problem*).

Key Learning Points for the Safety Leader

The place that information relating to health and safety at work occupies within the individual's paradigms will determine how it is mobilized and integrated (or not) into the individual's reasoning.

Bounded Imagination

We can see that our thinking is based on our paradigms (chiefly) and on the additional information we have at our disposal (sometimes), but we never have all the data. We realize that certain elements are missing (the "unknowns"), and we can make assumptions or incorporate a degree of uncertainty into our reasoning.

But there are also elements that we do not know about and that we do not even know exist. These are the *unknown unknowns* made popular by former US Secretary of Defence Donald Rumsfeld, which can play a decisive role in the problem to be solved: we do not know how incomplete our models are. Daniel Kahneman (Kahneman, 2011) used the acronym WYSIATI (*what you see is all there is*) to illustrate that we operate under the illusion of knowing what are all the parameters that we need to be aware of when making our choices.

This illusion can lead us to make mistakes not only in technical terms but also in terms of organization, anticipation, and planning, and even in our interpersonal relationships.

We think we can foresee the consequences of each of our choices, in a logic of causality. But even if our logic were perfect and we knew the main parameters of the equation, we would have to reckon with the contingent and emergent nature of certain phenomena in a system that is always too intractable for us to control or even accurately define. Expertise, and notably that coming from experience, can certainly expand the limits of what we can imagine, but it will never allow us to envision every potential scenario in a complex situation – and it can inflate our expert bias as described above.

Despite this, we often believe that we can imagine everything that could happen and therefore plan the appropriate responses accordingly. This is not true: our imagination (from the Latin *imaginatio*, image or vision) cannot be truly original. Our prefrontal cortex calls on our memory and on the areas of our brain linked to perception (described above) to develop "new" images and concepts, which in reality are always based on what we already know.

FOCUS ON... THE PLANNING FALLACY

From the Sydney Opera House to California's High-Speed Rail, not to mention your latest DIY project, the examples are everywhere: we systematically underestimate the time needed to complete a task. The optimism bias (the benefits of which have been demonstrated for our psyche to avoid despair and depression in a world that is not always very benevolent) combines with the necessarily incomplete information available to us and our inability to imagine unforeseeable problems.

Megaproject expert Bent Flyvbjerg brilliantly illustrated this phenomenon with multiple examples (Flyvbjerg and Gardner, 2023), but, once more, it was Kahneman and Tversky who, by the end of the 1970s (Kahneman and Tversky, 1977), exposed what they called the **planning fallacy**. This bias affects us when we have good experience with similar tasks and even when we have already suffered delays with them.

These planning errors can become a source of stress and haste, which, as we have seen, can increase the risk of accidents. Furthermore, as Kahneman also proposed later with Lovallo (Lovallo and Kahneman, 2003), this planning fallacy leads us to systematically underestimate the *risks* associated with an operation or project that we are in the process of planning.

The good news is that this phenomenon only affects the individual or group directly involved in planning the activity. In general, outside observers are more factual, even pessimistic, when it comes to evaluating completion times and the risks involved. Hence the good practice of bringing in a truly independent outside eye (consultant, expert, or other) to give its opinion at critical stages of the planning process. Even if, as we have seen, the experts can also be wrong...

Key Learning Points for the Safety Leader

When we try to plan and anticipate the progress of a task or project, we systematically underestimate the difficulties, the time needed, and the risks involved.

Thus, it is often a good idea to call in individuals from outside the team in charge to provide a more detached view.

Epistemic Arrogance

But not all problems stem from a lack of knowledge. What matters most is having the right kind of knowledge. Aristotle's distinction between *practical wisdom* (or *phronêsis*, discussed above) and *theoretical wisdom* is noteworthy. Contrary to Plato, Aristotle stated that even if it is possible for young people to acquire the latter by studying, "(by doing so) they apparently do not attain practical wisdom. The reason is that practical wisdom is concerned with particulars as well (as with universals), and knowledge of particulars comes from experience".

Expert bias as described above is indeed more problematic among those whose expertise is mostly theoretical (limited to *epistêmê*). Nassim Taleb pointed the negative and sometimes even dramatic effects of what he calls **epistemic arrogance** and wrote: "our knowledge does grow, but it is threatened by greater increases in confidence, which make our increase in knowledge at the same time an increase in confusion, ignorance, and conceit" (Taleb, 2010).

DID YOU KNOW? DO NOT CONFUSE EDUCATION *WITH* EXPERTISE

In a pioneering study subsequently replicated dozens of times (Alpert and Raiffa, 1982), researchers interested in *calibration* (assessing probabilities in a context of uncertainty) asked students at Harvard Business School to estimate the value of unknowns such as the population of Rajasthan or the number of lovers that Catherine II of Russia had during her life. To be noted: they could give as wide a range as they wished within which they thought the value should lie with a 98% probability. The error rate was still 45%! Subsequent studies showed much lower error rates (and therefore less arrogance) among cleaners and drivers.

All these studies are a reminder of the relevance of the famous quote from Thomas Mann: "There are so many different kinds of stupidity, and cleverness [in the sense of purely theoretical knowledge, note from the author] is one of the worst".

Paradoxically, possessing more information is not always a good thing. In a famous study conducted in the 1970s by Paul Slovic – who often worked with Kahneman – a panel of selected handicappers (handpicked professional horse racing gamblers) were given a list of 88 variables concerning the horses to judge the outcome in 40 races. Each handicapper could choose 5 variables first, then 10, 20, and finally 40 variables. Average prediction accuracy remained the same, while their confidence level grew to almost double their actual win ratio (Slovic, 1973).

Key Learning Points for the Safety Leader

Advanced theoretical knowledge can lead individuals to rely excessively on mental models that they may have acquired or developed (and which we have discussed above), to overlook certain verifications in their intellectual process, or to underestimate the margins of error in their conclusions, leading them to what is known as **expert error**.

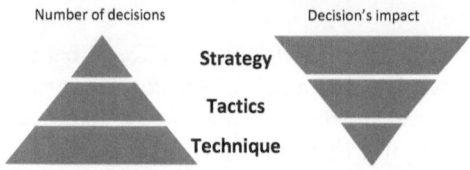

FIGURE 52   Strategical, tactical, and technical decisions.

STRATEGY, TACTICS, AND TECHNIQUES

Equipped with these imperfect knowledge, processes and tools underlying our logic, we have to deal with a wide range of risks (and opportunities). Different types of decisions will have to be taken: techniques, tactics, and strategies will have to be defined and implemented.

The distribution of the number of these decisions can be illustrated by a pyramid, which reminds us of the pyramid-shaped org-chart of most organizations: strategic decisions would be the prerogative of executives, tactical decisions the domain of management and technical decisions the domain of engineers and other "technicians", while operators would be content to carry out the work (Figure 52).

The reality is rarely so simple: managers and even executives are often very busy – sometimes too much – with purely operational issues, activities that remain nevertheless necessary to develop and maintain their *phronêsis*. And all the players in a team are thinking strategically, more or less deliberately and elaborately, about their role and their interactions with the organization and with others.

Technical Thinking

In the workplace, most of the problems that arise on a daily basis, and which can have an impact on health and safety, are **technical** in nature. Please note that in this book we do not understand the technique to be restricted solely to technology, but we use the term in a much broader sense, closer to its etymological roots (*techné* is ancient Greek for making, doing, or effective action, and is counted by Aristotle as "practical knowledge" among the five human intellectual virtues that he described). It refers in particular to the skills, know-how, and methods deployed by workers, whether they involve the use of tools and other instruments or just their own bodies.

Technique therefore concerns everything to do with the execution of tasks, both by humans and by the machines they control. Indeed, at this level, human beings as operators could be compared to (if not replaced by) the automaton in that they do not have to worry about anything other than carrying out the task entrusted to them. The issues that may arise at this level would be related to **how to do** the task, for the sake of **effectiveness** in the sense of the first part of ISO's definition for this concept: *the extent to which planned activities are realized.*

In tractable systems, we follow precise procedures and stick to detailed execution instructions that are defined to remove as far as possible any operational ambiguity. They can become very tightly coupled. Some of these technical constraints have become regulatory obligations and/or are the subject of standards issued by national and international standardization bodies: NFPA standards for fire safety, CEN-CENELEC standards for electronics and electrical engineering, etc. In intractable (or simply less regulated) systems we rely mainly on craftsmanship and expertise.

A priori, our techniques and the technological systems that derive from them and support them are supposed to function correctly or at least with a sufficient degree of reliability, provided that they are used and maintained properly. However, as we have seen so far, there are many limitations that can hinder the proper execution of a task by humans and cause accidents, and which must be taken into account at the *tactical* and *strategic* levels.

Tactical Thinking

Even if the details of these human and technical limits may be unknown to some, in absolute terms nobody expects perfect technical mastery leading to infallible performance. Still, a priori, this should not cause serious accidents: our **tactical thinking** should anticipate components' and subsystems' failures that can pose problems for the individual, the teams, and the machines during the execution of the activities.

At the tactical level, we must decide on the techniques to be used, and therefore the human and material resources to be deployed, to achieve a successful result. We need to decide **what to do**, with a view to **efficiency**, defined by ISO as the *relationship between the result achieved and the resources used.*

This reflection must therefore lead us to arbitrate between different operational priorities, decide on the nature of the actions and controls, plan their sequencing, and ensure that the necessary resources are available. We define procedures and protocols, and we manage projects and programmes.

This is what we commonly call *management*, and so it is traditionally the job of managers to look after it. They can also rely on standards, this time for management systems (such as ISO 45001), which are necessarily more generic than technical standards since they specify the *what* and not the *how*.

As we saw earlier, most workers today are *knowledge workers*, occupying roles in which a certain amount of tactical thinking is imperative. However, this operational need for tactical thinking is very rarely formalized or recognized, which can create different types of risk situations.

On the one hand, workers regularly need to make tactical choices (for example should I follow the lone worker SOP or not, given that my colleague will be away just a few moments?) whereas they may lack the expertise required to answer the question and even the necessary perspective required to raise the question in the first place.

On the other hand, in a highly regulated work environment, an experienced worker may have the competence to appreciate the need for a tactical change (for example switching to a degraded mode operating protocol in the face of an unforeseen change in operational parameters) but not have the autonomy or authority to take this kind of decision.

Strategical Thinking

In the end, the **strategical decisions** are the ones that have the greatest impact at both the individual and organizational levels, including in the field of safety. These are decisions that are taken with an eye to the famous big picture, the long term, and the vision, mission, purpose, and values of the organization, including the value it places on safety. We will be coming back to this notion at length in a moment.

It is in this light that we define, at best as clearly and explicitly as possible, at worst in a deliberately or unintentionally ambiguous way, the doctrine and policies that orient, guide, and provide a general framework for action. At this level, we think about **why**, in the interest of **efficacy** or **effectiveness** but related to the second part of ISO's definition: the extent to which *planned results are achieved.*

If tactical thinking allows us to draw the roadmap, strategic thinking sets the goal and provides us with the right compass. Because, as Peter Senge said, you can move through the jungle very efficiently while moving in the wrong direction or even in the wrong jungle (Senge, 2006).

Former President of WHO's European Environment and Health Committee, William Dab, proposes a method comprising five questions to ask yourself when setting up a prevention programme, to maximize the chances of success. Answering "no" to any of these questions should lead the decision-maker to look for additional data (Dab, 2018):

- Do we know the scale of the problem and the magnitude of the risk?
- Do we have a good understanding of this risk and the factors involved?
- Are intervention tools available, and do we know how effective they are?
- What are the conditions for the feasibility of the intervention? Are they in place?
- How will the success of the intervention be assessed?

Many executives take pride in their pragmatism and their "on the ground" approach to safety and demand the same from their supervisors and safety officers, as well as from the entire management team. It is undeniable that technical and tactical aspects must be dealt with, that prevention requires *phronêsis*, and that it involves working on the front line to resolve operational problems.

But just as these executives would not think of launching a new product without a business plan (based on a commercial strategy) or building a production line without calculating its return on investment (based on an industrial strategy), it seems ill-advised to neglect strategic thinking when it comes to occupational health and safety. Sometimes we need to take a step back and get some perspective.

In addition to guiding and framing prevention initiatives and determining the resources to be devoted to them, whether financial, organizational, or human (the tactical dimension), this strategic reflection must review the beliefs and values on which these initiatives will be based. It must address the **culture** of the organization and the place where safety should be within it. Once this *ethical framework* has been established, it will be possible to ensure that the work carried out in the field and the tactics and techniques adopted also help to demonstrate the commitment and value that the organization and each of its leaders place on safety.

These concepts will be analysed in detail in the following sections because, far from being anecdotal, they are the main levers available to leaders to influence how their teams prioritize prevention on a day-to-day basis.

DID YOU KNOW? EISENHOWER'S MATRIX

President Dwight D. Eisenhower would have said that "what is important is rarely urgent, and what is urgent is rarely important". It was on the basis of this quote that the matrix below was proposed (Figure 53).

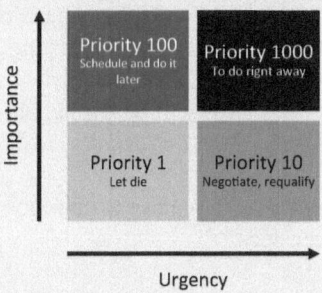

FIGURE 53   Eisenhower's matrix.

It is a good reminder for leaders who tend to get sucked in by emergencies at the cost of always leaving the fundamental, structuring, and truly transformative issues for later. Safety improvement programmes are a perfect example of an important subject that, in the absence of any particular urgency (no serious accident, for example), "falls by the wayside" both in the agendas of management committees (always too full) and in budgets (always too tight).

Key Learning Points for the Safety Leader

Whatever our position in our organization, we spend most of our time dealing with operational, technical, and tactical issues, and this is also true in the field of occupational health and safety.

But safety leaders, whatever their hierarchical level and function, must also have a strategic vision of these issues to find the most effective ways of shaping the organization's culture and thus orienting attitudes and therefore all decisions, choices, and behaviour towards safety. We shall see how shortly.

NOISE

One of the most insidious limitations of our reflective processes is the dispersion of their results, already at the level of the individual and a fortiori between different individuals. We know that our **judgement**, understood as the ability to use our deliberative capacities described above to estimate measurements and form projections, is not perfect – but we are generally far from being aware of the extent of its unreliability.

Usually, we set or "freeze" our judgements and predictions the moment we pick up an internal signal, a sort of self-generated reward that satisfies us intellectually when we estimate that we have put together all the data in our possession into a coherent narrative. However, such satisfaction does not objectively imply any guarantee of validity (we will talk more about these signals when we deal with affects and emotions).

Daniel Kahneman, this time with Cass Sunstein (who coauthored the popular *Nudge* with Richard Thaler) and Olivier Sibony, have recently studied this dispersion, which they call **noise,** and which must be distinguished from the biases presented above. Bias will push our judgement in a particular direction, while noise (or dispersion) means that our judgement will be wrong, but we cannot predict in which direction or to what magnitude. Bias and noise add up as independent sources of error (Kahneman et al., 2021).

The two things that biases and noise have in common are their ubiquity and their potentially negative consequences.

FOCUS ON... DIFFERENT TYPES OF NOISE

Within a system, noise can be broken down into *level noise* and *pattern noise.*

Level noise refers to the variability between the average judgements of different individuals: for example, some magistrates are harsher than others, with a higher average sentence. In occupational safety, we might think of a worker who has a general propensity to take more risks than her colleagues.

Pattern noise, quantitatively more important and more difficult to counter, refers to the variability between the specific judgements of two different individuals in relation to the same instance: for example the difference between the sentences that two judges would assign for the same crime. In the safety domain, we can imagine a worker who pays little attention to traffic risks (even though he systematically wears his PPE) and the opposite case of another worker who pays a great deal of attention to traffic (but neglects to wear his PPE).

In turn, a sub-component known as **occasion noise** can be observed in level noise. It corresponds to the differences between judgements made by the same individual in front of identical instances but at different times: for example, a judge who would be more lenient the day after the victory of the team of which she is a fan. In the occupational context, we can think of workers who take more shortcuts when tired or hungry or managers who are less careful about these shortcuts at the end of the month when it is time to "get production out the door".

With the exception of occasion noise, noise can be attributed to the different **attitudes** (and related values and beliefs) of different individuals as we will define them.

These authors acknowledge but do not address the issue of values and beliefs and limit their recommendations to a series of management techniques, including recourse to a second opinion or the application of rules defined on the basis of statistical analysis of past occurrences. As a matter of fact, a number of studies have demonstrated the overall superiority of rules (such as simple decision algorithms) over case-by-case judgements and assessments, even when considerable expertise is involved. But we have seen that not everything can be effectively regulated.

DID YOU KNOW? VERY NOISY SENTENCES

In the 1970s, the American judge Marvin Frankel drew attention to the wide disparities in sentences handed down in the United States for comparable offences. Studies were carried out, including one in 1981 in which 208 federal judges were asked to rule on 16 hypothetical cases. In only 3 of the 16 cases, there was unanimity on the imposition of a prison sentence.

Even when the majority of judges considered a prison sentence to be appropriate, the length of sentences varied substantially. For example in a case where the average prison sentence handed down was 1.1 years, the longest sentence handed down was 15 years (Kahneman et al., 2021).

Key Learning Points for the Safety Leader

The problem of the dispersion of judgements (and the behaviour that follows) that results from the limits of our rationality that we have just studied can legitimately be approached from the point of view of rules in the context of a regulated, tractable system.

But given the intractable nature of many work situations, in the next section we will begin to explore a different and complementary approach: attitudes, based on the individual's values and beliefs.

TOOLBOX – Tool No. 5: Mnemonic, Judgement, and Reasoning Nudges

Richard Thaler and Cass Sunstein propose an approach to influencing behaviour that they call **libertarian paternalism** (Thaler and Sunstein, 2008), which consists of trying to guide people's behaviour by using the automatisms, biases, and heuristics to which they are usually subject in both their System 1 and their System 2. The approach would be *paternalistic*, accepting that individuals need to be guided to make the best choices in their own interests, but at the same time *libertarian*, as they would retain freedom of action.

Based on this principle, they consider that those in charge of a system must organize the context in which individuals make their decisions to orient them, and they call this discipline **choice architecture**, nicknaming it **nudge**, the name under which it has become popular.

This approach is highly attractive in cultures where individual freedom is a central value, such as the United States or the United Kingdom, where David

Cameron's government set up the Nudge Unit within the Cabinet Office in 2010, with the very official aim of promoting the application of behavioural sciences in all the country's public policies.

A series of techniques uses this approach to influence spontaneous behaviours linked to System 1, starting with sensorimotor and attentional biases. The first step is to identify and eradicate any stimuli that could be a source of error (for example a sign at the bottom of a staircase that attracts the attention of someone coming down the stairs, diverting it away from the steps). Then we look for and implement devices carrying the stimuli that will trigger the expected behaviour (for example painting the edges of the steps with a high-contrast colour to attract attention) (Figure 54). The aim is to maximize *stimulus–response compatibility.*

FIGURE 54   Staircase (picture by author).

These techniques are based on the principles of Watson's respondent behaviours presented earlier, which rely on external stimuli in the form of sensory signals, mainly visual, but also auditory and even olfactory. On top of the examples presented in Tool No. 2, we can mention:

- Deterrent lines between lanes on motorways in France, with lines (always 3 m long) that grow closer together around entrances and exits (with intervals of 1.33 m, instead of the usual 10 m), designed to discourage people from changing lanes at these points.
- The addition of odoriferous molecules (mercaptans) to natural gas gives it a characteristic unpleasant odour.

Other techniques aim to influence deliberate behaviour linked to System 2. Choice architects will try to organize the information available to the individuals to guide their choice while using their own memory, judgement, and reasoning biases. Examples of the use of these biases to encourage the adoption of behaviour include:

- Linking the behaviour with objects, people, or qualities generally appreciated by the target individuals (e.g. providing young workers with safety shoes with a sporty design).
- Presenting the behaviour as being widespread within a group or category of people (e.g. publicizing that 98% of employees wear their safety glasses).
- The use of default choices, with most people not bothering to consider the alternatives and keeping the default choice even if they have the opportunity to change it (e.g. choice of health insurance options).

Attentional, memory, judgement, and reasoning biases operate in a given cultural context, and it is therefore necessary to study this cultural dimension to understand their effects and find ways of acting on them (Blanco-Munoz, 2021).

ATTITUDES

"Attitude is the mind's paintbrush. It can colour any situation".

Alexander Lockhart

Despite the difficulties in measuring them and some studies pointing to inconsistencies between *declared* attitudes and behaviours actually adopted (Wicker, 1969), the concept of **attitude** as the association learned by an individual between an object (material or immaterial, such as behaviour) and the positive or negative evaluation she makes of that object is not only scientifically valid but also constitutes a very good tool for analysing behaviour and building strategies to influence it in its social context.

We understand attitudes as **behavioural preferences**: a *tendency* to adopt a given behaviour, similar to what Immanuel Kant called *inclination* – the preference for some states of the world over others as determined by what one finds *pleasurable* – and that he considered to be the basic explanation of human behaviour. Let us already note that Kant also pointed that people can also act from *duty*, being moved to act just by the recognition that an act is morally required, according to a notion of ethics that we will get to later.

What was once a purely philosophical concept has now gained neuroscientific support. Neurobiologist Antonio Damasio theorizes that our experiences establish neural connections between categories of events and objects and positive or negative emotions, which he calls **somatic markers** and which we will describe a little later masio (Damasio, 1994). Influencing both automatisms and deliberate thought, our attitudes have a subconscious influence on almost all our behaviours (Fazio, 1990).

But it is not only external stimuli, charged with meaning through direct experience of the consequences, which originate and consolidate our attitudes. As proposed by renowned psychologist Albert Bandura, *social learning* builds up complex associations internalized by the individual, which we are going to dissect here to better understand them and be able to guide them (Bandura, 1986, 1999).

DID YOU KNOW? ATTITUDE, CHARACTER, AND PERSONALITY

Attitude, character, and personality are concepts that have been historically confusing and ambiguous. Aristotle referred to *character* as a set of principles that we acquire through "sheer force of will". More recently, in his famous book *The Seven Habits of Highly Effective People*, Stephen Covey employs the term *character* in a sense very similar to the one we give here to *attitudes*, and gives to the term attitude a meaning quite different from the one we are giving it in these pages (Covey, 1989).

Importantly, Covey makes a crucial distinction between this concept, which we shall call attitudes, and that of *personality*. Whereas – in the sense we give to it – attitude is specific to a set of behaviours, personality is a vaguer and more ambiguous concept. It has been studied by psychologists through many different lenses, including biological, cognitive, and trait-based theories and learning, as well as psychodynamic and humanistic approaches. The question of to what extent personality is originated by Nature and how much of it comes from nurture is still very much open.

Personality will without any doubt orient the individuals' behaviour, for better or for worse, in the occupational health and safety domain. But although personality definitely evolves and matures throughout life, we have little hope to alter it by any means at our disposal in the workplace. Attitudes, on the contrary, we can influence, and we shall see how.

These associations and the resulting behavioural preferences that we call attitudes are presented by psychologists as constructs built on three dimensions: *cognitive*, *affective*, and *conative* (Ajzen and Fishbein, 2005). This conative dimension, which inhibits or mobilizes the will to act and then orients the action or choices, is thus inextricably integrated in the frame of our attitudes with our cognition and our affects.

Behavioural preferences can therefore be influenced by interventions targeting the cognitive and affective dimensions of the individual, through his **beliefs** and **values** by shaping, as we will see, the cultural context in which they are acquired (Hofstede et al., 2010).

Once developed in the individuals, these attitudes are generally fairly stable and will *guide* their behaviour, but will not *determine* it. The individuals' behaviours will also be influenced in a given situation by their perception of their environment, a subject covered previously. These three components – perception, values, and beliefs – are interdependent and continually interact with each other (Figure 55):

- Daily exposure to certain cultural elements or *artefacts* (a notion that will be further developed later), which shape the individual's values and beliefs over the long term.
- In turn, values and beliefs will influence what the workers perceive, what will become cognitively salient to them, and how they interpret it (as we saw previously).
- Values and beliefs influence each other, sometimes to the point of being inextricably linked. For example, if one believes that neither a person's gender nor colour a priori determines their skills or character, one will be attached to the values of equal opportunities and fair treatment. And the other way around, if we cherish values such as equality and fraternity, we will find it difficult to accept reasonings supporting discriminatory practices.
- Finally, the individual's self-perception of oneself executing the "expected" behaviour in line with the values, beliefs, and related artefacts to which one is exposed will reinforce their appropriation.

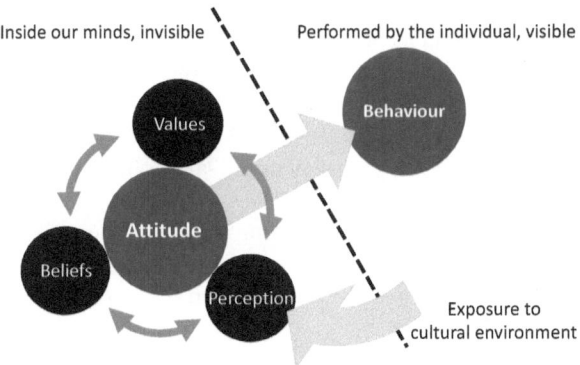

FIGURE 55   From attitude to behaviour (drawing by author).

The celebrated Stanford professor James G. March, famous for his work on organizational decision-making, explained in the 1950s that, rather than being based on rational choices involving preferences and expected consequences, decisions were made on the basis of rules that are appropriate to the situation and the *identity* of the decision-maker (March, 1958). This notion of identity can be associated with the concept of attitudes and the underlying values and beliefs.

An attitude may be considered "adequate" or "satisfactory" in relation to safety in a certain context insofar as it raises the workers' situational awareness and orients them towards the execution of safe behaviour. On the contrary, it may be deemed as "unsuitable" or "inappropriate" in relation to safety if it may lead the individual to ignore dangers (consciously or subconsciously), take shortcuts, or fail to ask questions or raise concerns.

Another ABC model (different from the one proposed by behaviourists and presented above) is formulated as follows:

Attitude → Behaviour → Consequence

FOCUS ON... THE MODE MODEL

Considered by neoclassical economists, as preferences, as mere inputs to a rational calculation, and ignored by behavioural psychology, attitudes have nonetheless been at the heart of social psychology studies for almost a century.

Although the concept of attitude has been theorized for such a long time, it has long been somehow neglected by the science of behaviour in general and behavioural safety in particular, at least in its experimental aspects (Wiker, 1969). Because it is not possible to directly observe and measure someone's attitude towards an object or behaviour.

Therefore, researchers wishing to validate the various theories on behaviour, and safety practitioners developing methods to influence them in the field of risk management, have traditionally focused on setting up experiments and tools aimed directly at the phenomenon they were interested in – behaviour – without digging any deeper to find its origins.

Even though scientific literature has finally emerged on the subject and attitudes are now considered by many to be a valid key to most of our behaviours (Homer and Kahle, 1988; Fazio, 1990), this concept remains blurry in most practical methods dealing with behavioural safety, human factors, or safety culture (Cooper, 2001).

And it is rather paradoxical that for almost 40 years there has been so much interest in the notion of safety culture, without this concept of attitudes, which lies precisely at the junction between culture and behaviour, having been given greater prominence.

This apparent lack of interest can be explained at least in part by the lack of a clear definition of the concept, its neurological basis, and its psychological implications. It is not that the concept was strange to occupational safety practitioners, but it has been largely misunderstood and therefore underutilized.

We will retain the MODE model, with which Russell Fazio (Fazio, 1990) explains how these attitudes influence both spontaneous (System 1) and deliberated behaviours (System 2). Figure 56 illustrates these mechanisms:

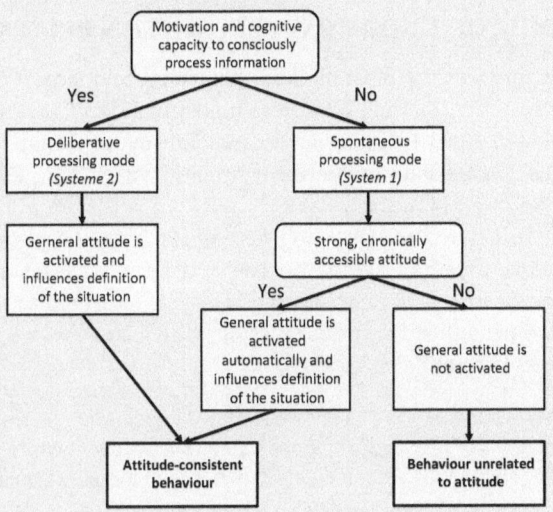

FIGURE 56 The MODE model. (Adapted from Fazio (1990))

Importantly, the intensity of the association – and therefore of the attitude – determines not only the probability of its activation and the degree of influence on the triggering of the corresponding behaviour: it also determines its capacity to bias the individual's perception and interpretation of new information, and therefore their resistance to changing their attitude.

BELIEFS

The first pillar of our attitudes that is fashioned by our day-to-day perception, the notion of **belief**, echoes that of "paradigm" presented above. These are deep-seated convictions about what is true and what is false, what is possible or probable and what is not, what works and what does not... that we integrate consciously or subconsciously as a certainty into our mental models.

Our beliefs will obviously weight on our "rational choices" when we ponder the pros and the cons of different alternatives. The way they also influence our spontaneous, "system 1" decisions may seem more obscure. We can link our beliefs to Klein's "expectancies" in his Recognition-Primed Decision Model described earlier. Once the familiar situation has been recognized as typical, we project into it our mental models and all the beliefs that come with them.

We acquire and consolidate our beliefs through experience within the limits of what we can personally do and be exposed to, as well as through social learning or *acculturation* within the various moral circles to which we belong throughout our lives (family, community, country, religion, training, profession, company, association, etc.). We will come back to this cultural dimension later, because it is at this level that we must act if we hope to orient the beliefs of the members of a group.

DID YOU KNOW? CAPTAIN SMITH'S BELIEFS

The RMS Titanic was the most modern, luxurious, and largest ocean liner in the world at the time, standing 25 storeys high and 275 m (300 yards) long. On the night of 14–15 April 1912, during her maiden voyage from Southampton to New York, the Titanic sank in the North Atlantic with 2,227 people on board, taking the lives of 1,514 of them.

Sailing at full speed in dense fog notwithstanding the iceberg warnings from other boats in nearby waters was one of the direct causes of the tragedy. Below is an extract from an interview given by Edward J. Smith, captain of the RMS Titanic during this disastrous crossing, with a New York Times reporter in 1907 (5 years before the sinking):

> "When anyone asks me how I can best describe my experience in nearly 40 years at sea, I merely say uneventful. Of course there have been winter storms and gales and fog and the like, but in all my experience I have never been in an accident of any sort worth speaking about".

Smith was also quoted in 'The World's Work' in 1909, referring to the Adriatic, his ship at the time:

> "I will not assert that she is unsinkable, but I can say confidently that, whatever the accident, this vessel would not go down before time had been given to save the life of every person on board. I will go a bit further. I will say that I cannot imagine any condition that would cause the Adriatic to founder. I cannot conceive of any fatal disaster happening to this ship. Modern shipbuilding has reduced that danger to a minimum".

Examples of prevention-related beliefs:

Positively:

- It's better to take your time than risk an accident
- If in doubt, always check
- We're all fallible, so we need to take precautions
- Wearing a helmet has already saved more than one life
- …

Negatively:

- This machine/equipment/process… is totally safe
- I know it can be dangerous, but it's OK: I know what I'm doing/I'm careful
- Everybody does it/We have always done it like this
- We can't do it any other way: it would be too long/too tedious/too expensive
- …

The classic strategy for trying to change people's attitudes and, ultimately, their behaviour is to try to change their beliefs by implementing **fact-based communication initiatives**. Workers are presented with dangers, accident reports, statistics, etc., in a factual and objective way, in the hope that this will enlighten them and make them adopt safe behaviour and avoid risks.

We have all seen the effects, modest at best, of most communication campaigns of this kind. Climate change, smoking, healthy lifestyles, and road safety are sadly edifying examples.

When these messages diverge from the individual's prior beliefs, the effects of this communication will be moderated or even cancelled out by the confirmation bias described above. Worse still, we may trigger mocking, cynical, or even openly hostile reactions from the individual. Indeed, challenging deeply ingrained beliefs amounts to challenging the individual's self and identity, and they naturally react in a defensive way. Certain tactics, such as paradoxical communication, can help us to overcome these difficulties (see Toolbox, Tool No. 6).

But in all cases, a distinction needs to be made between informing and communicating (we also sometimes talk about raising awareness in the sense of influencing attitudes, but that is the common objective, not the approach). **Information** goes one-way: the sender expects her audience to receive a message and hopes that it will be understood and accepted. **Communication** involves dialogue and sharing points of view, with a right to reply that encourages exchange and discussion in a climate of trust and psychological safety, concepts that we will develop later.

Information campaigns such as posters, mailings, and other multimedia tools are undoubtedly a useful *complement*, as a reminder, but cannot have the same impact as genuine dialogue. What is more, if disconnected from the organization's culture as we weill describe it, posters and other videos played over and over again on screens quickly become "part of the *décor*" (as per the habituation phenomenon already described) and are therefore completely useless, and even counterproductive, as yellowed and dusty evidence of management's lack of genuine interest in safety.

Key Learning Points for the Safety Leader

The individuals' beliefs play a key role in their assessment of the risks that they will take, the efficacy of the control means deployed, and the subjective feeling of safety or danger that they will experience, thus influencing their behaviour. These beliefs are generally well-established and cannot easily be changed simply by displaying the information we want them to integrate.

Every leader needs to be aware that, if they really want to change the beliefs of their teams, propaganda tools are not enough: they need to speak out and address them directly to explain, listen, and respond. Leaders at the top of an organization need to design, implement, and maintain management routines that ensure this capillary communication on a daily basis in a climate of trust.

TOOLBOX – Tool No. 6: Paradoxical Thinking

We have seen how confirmation bias means that we readily integrate information that supports our beliefs, but we are not very receptive (or totally impervious and sometimes even hostile) to information that goes against them. Here is an innovative approach aimed specifically at influencing the attitudes of people whose beliefs are far removed from those we might expect, and who are therefore "resistant" to conventional messages.

Researchers at Tel Aviv University experimented with the use of **paradoxical thinking** to influence the extremist attitudes of certain people towards the conflict with Palestine (Hameiri et al., 2019). After a pilot carried out in 2013 with 161 subjects, in 2015 they conducted a full-scale experiment in a small town in central Israel, where 63% of the population voted for parties supporting aggressive policies (the *hawks*).

For 6 weeks, they displayed posters, brochures, balloons, t-shirts, and other publicity items in the streets, as well as Internet banners targeting IP addresses in the area, with messages such as the following:

- "Without it, we would never be just: to have justice, we probably need conflict".
- "For the heroes, we probably need conflict"

Of course, no one in this population was aware that they were part of a psychological study.

6 weeks after the campaign, they carried out surveys on this community to compare it with a similar control group that had not been exposed to these messages. They found that perceptions of conflict and attitudes towards aggressive policies had changed among those exposed, but *only among the most extremist*. These showed a significant decrease in their attitudes in support of the conflict (Figure 57).

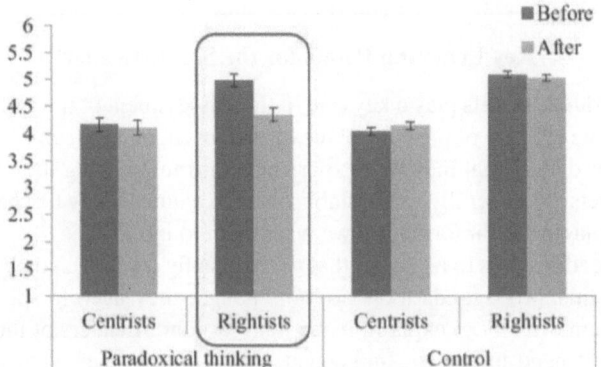

FIGURE 57 Evolution of attitudes after intervention vs control group (Hameiri et al., 2019).

They proved that you can effectively influence extremist attitudes in individuals by subtly reducing the underlying beliefs to absurdities, using messages that are:

- In line with their beliefs
- But significantly more extreme than those they would hold.

By using this tactic, the individuals targeted do not adopt a defensive or rejectionist stance but are instead challenged and made aware of the inconsistencies in their beliefs.

This same technique has been used by this author in an industrial context to target the workers most resistant to conventional prevention messages. By highlighting the absurdity of certain attitudes and behaviours, we get them to question themselves, which is more effective than presenting them with ideas they are not prepared to listen to (Figure 58).

"This way I save time"   *"Hurry up, don't be late!"*

FIGURE 58 Examples of posters used for a paradoxical thinking intervention (pictures by author).

Preparation, including prior notification of management and staff representative bodies, is of course necessary, as well as debriefing of workers by field management after exposure of these messages to complete the communication.

VALUES

The second pillar on which we build our attitudes is our **values.** These are general tendencies to prefer certain states, worldviews, and ideals over others, which make us assign things, facts, or people with a positive or negative *appreciation* (or *emotional valence*). As cognitive as well as affective representations of *desirable, abstract, and transsituational goals* (Rokeach, 1973; Schwartz, 1992), it is our values that confer *importance* to the objects (material or intangible) on which we project them and serve us as guiding principles in our lives.

Like our beliefs, they are essentially transmitted to us by social learning, the education we receive in the *moral circles* to which we belong, education which is not always formal but more often the result of exposure to (and then appropriation of) practices, gestures, and other artefacts imbued with meaning, as we will see.

As we have already mentioned, beliefs and values are intimately linked: the former set the cognitive framework for our attitudes, and the latter create the *affective framework,* both frameworks being inextricably intertwined.

Faced with situations or behaviours that run counter to our attitudes, we will experience a psychological tension or *cognitive dissonance* (Festinger, 1985) that we will try to resolve. However, if exposure to situations or behaviours that go against our purely intellectual beliefs challenges us and can provoke a certain amount of conflict, this can be resolved by rational adjustments (starting with simple denial). But, situations or behaviours that go against our values (and the beliefs that are closely interwoven with them) elicit much more intense emotions of indignation and even anger and outrage.

FOCUS ON... THE SOMATIC MARKERS HYPOTHESIS

The work of neurobiologist Antonio Damasio has changed the way we look at the emotional component in our mental processes and choices. His studies of the brain led him to identify the regions involved in the representation of emotions, the lesion of which impairs the individual's ability to plan logically and make rational decisions, thus preventing them from behaving "reasonably" (Damasio, 1994).

Following on from this research, he postulated that genetically pre-set mechanisms for the fundamental homeostatic regulation of the organism serve as a basis for somatic appreciation of the "pleasant" or "unpleasant" nature of the circumstances affecting it (for example detection of low blood glucose levels) and trigger instinctive adjustment reactions to adjust them (in this example: looking for food). These are the mechanisms underlying our *primary emotions.*

We share these mechanisms with other species, but the superior mnemonic (memory) capacities of human beings mean that this initial basic repertoire is rapidly enriched by experience: new images, scenarios, or *potential representations* will be coupled to those linked to innate regulation by association via modifications to the neural space, in the form of what Damasio calls *topographically organized representations* in different areas of our brain, particularly in the prefrontal region. These are the mechanisms underlying *secondary emotions*, which we can identify with our **values**, which are juxtaposed with genetically pre-set potential representations that give them a somatic appreciation.

How do these secondary emotions affect our cognitive processes? Damasio proposes that these representations of a state of the body and its positive or negative appreciation constitute **somatic markers,** which can come into play in two ways following the perception of an object or situation (or its mental evocation):

- In the *basic mechanism*, the prefrontal cortex and the amygdala establish a particular profile of the state of the body, which triggers visceral signals that are then picked up by the somatosensory cortex.
- In the *alternative mechanism*, the prefrontal cortex and the amygdala send a simulation of a body state directly to the somatosensory cortex. These simulation mechanisms are thought to develop with learning, as the individual is able to classify repeated situations into categories until they take precedence over the basic mechanism.

In both cases, the somatic states picked up by the somatosensory cortex will influence perception, direct our attention, and affect our reasoning, even to the point of subconsciously ruling out alternatives linked to negative representations.

It is in this way that our cognitive processes would be indissolubly linked to emotions.

Here we see that values are very powerful *motivating* factors (Rohan, 2000; Seligman et al., 1996), and we will come back to these motivational aspects later. For the moment, we can already observe that acting in accordance with our values represents a reward in itself, a source of well-being, whereas acting against our values provokes unpleasant emotions, an internal conflict that we will find difficult to alleviate with rational justifications Importantly, emotion and affects operate more often than not below the radar of our awareness and influence our cognitive processes and behavior in an unconscious way. The do not equate *feelings*, which are the small tip of the emotional iceberg that we are consciously aware of.

When we perceive a potentially dangerous situation and "lean", consciously or subconsciously, towards one behaviour rather than another, we can imagine each of these options as being supported by underlying beliefs and values. Sometimes we may be faced with contradictory beliefs, tensions that can be adjusted through reflection. But very frequently we will find there values competing with safety: speed or simplicity (and therefore a certain vision of efficiency), comfort, etc.

The image of the cartoon character torn between two choices, with a little angel on one shoulder and a little demon on the other, each telling her respectively to make the "right choice" or to allow herself to be tempted by something that attracts her but that she recognizes as being "wrong", is a pretty good representation of this process, which most of the time takes place at a subconscious level (Figure 59).

FIGURE 59   Competing values (drawing by author).

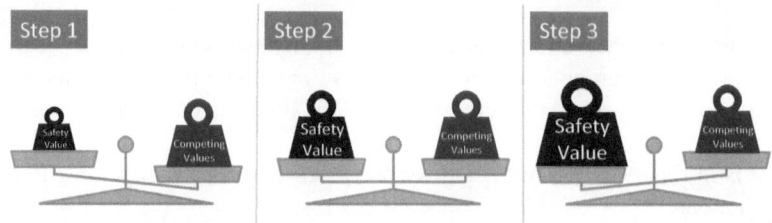

FIGURE 60   Evolution of values (drawing by author).

In these circumstances, it is the value that carries the "most weight" for us that will tip the balance and decide which attitude will prevail. As we shall see, the main role of leadership is to continually reinforce the value placed on safety.

In this light, the PIC-NIC technique already presented takes on another dimension that goes well beyond behaviourism: the immediate and certain feedback on a daily basis will also and above all slowly but surely reinforce the weight that safety has as a value in workers' attitudes (Figure 60).

This is a very gradual process, more or less difficult depending on the individual's various values, their compatibility with safety, and their intensity. Schwartz listed ten basic values, among which we find security – which we can assimilate to safety. The other nine (conformity, tradition, power, achievement, hedonism, stimulation, self-direction, universalism, and benevolence) can either be aligned with safety and add weight to the cause of prevention at the time of making a choice or, on the contrary, motivate us to adopt an unsafe behaviour.

We will see later what other tools leaders have at their disposal to accomplish this complex task, but to complete what has been said about communication to influence beliefs, we can already look at the impact that communication can be expected to have on values.

We can imagine and deploy campaigns pushing messages targeting the workers at an affective level. Even if, as with factual information, we can only expect from these other campaigns at best a reinforcing or reminder effect that will quickly become obsolete through habituation, they have often proven to be more impactful and effective than those aimed at "convincing" and changing beliefs by confining our speech to facts and figures.

These sequences, which are designed to appeal to emotion, work on two levels:

1. Leveraging values that are already present in the worker, i.e. abstract goals that are important or dear to her, by projecting them onto safety: family, hobbies, appreciation by management and/or peers, etc.
2. Developing safety's intrinsic value as an object that is in itself important for the collective and must be so for each of the individuals within it (Kahle, 1983; Homer and Kahle, 1988; Schwartz, 2012).

Examples of values linked to prevention:

- **First and foremost, safety as a value in itself: the importance of life, the avoidance of suffering, and the physical and psychological integrity of each individual.**

Other values to which prevention can be linked:
- The family, the loved ones we want to see when we get home from work and that we want to spare from worry
- My health, my ability to do other activities (my favourite sport, travelling...)
- Respect for others and rules
- Appreciation/validation from my superiors/peers/co-workers (in a context where safety is a shared value)
- ...

Values that may compete with prevention:
- Autonomy, independence, individual freedom ...
- Personal comfort
- A certain distorted vision of efficiency, time-saving, or economy
- Appreciation/validation from my superiors/peers/co-workers (in a context that is not conducive to safety)
- ...

DID YOU KNOW? SOCIAL NORMS VS MARKET NORMS

Perhaps the most common criticism of the values-based approach to prevention is: "That's all well and good, but people come here to earn a living, and all they care about is money". McGregor's Theory X has a very thick skin...

On the basis of this *belief*, which is of course valid in a professional context with often complicated employer–employee relationships, many organizations have introduced bonuses and other incentives (reserved for management or extended to all employees, either individually or collectively) as soon as a certain "safety objective" has been reached, generally a target injury frequency rate.

The ultimate effectiveness of this bonus approach is dubious for several reasons: for example these schemes can generate feelings of injustice ("such or such accident was sheer bad luck and should not be accounted for"), iniquity ("why should we penalize everybody just because that person did something stupid?"), double punishment ("our working conditions are harsh, and on top of that, if we get injured, we lose our bonus"), or partiality (temporary workers and contractors are rarely concerned by such bonuses, even though they often perform the hardest and most dangerous jobs). And that is without mentioning the deleterious effects on transparency.

In the absence of studies confirming or refuting the positive effect of this type of financial incentive on safety at work, everyone is entitled to their own opinion (their belief) on the subject...

But we can look at the experiment carried out in 1998 by a group of Israeli researchers over a 20-week period in ten kindergartens in the city of Haifa (Gneezy and Rustichini, 2000). They introduced a system of fines for parents who arrived late to collect their children at the end of the day. Classical behaviourism and economics assume that these financial penalties would reduce lateness. In fact, the opposite was true: the number of late arrivals increased significantly!

As soon as parents realized that they were in fact paying for the extra child-care time, they stopped feeling guilty for being late. In short, researchers found that the **social norm** (respect for others by being punctual) was a more powerful motivating factor for being on time than the **market norm** (the cost of the fine).

As Dan Ariely observed, when you offer someone financial compensation in a situation that is normally governed by social rules, the payment actually reduces their motivation to act (Ariely, 2010).

This example shows how values are a genuinely effective lever for guiding behaviour, even more so than mere financial incentives. In fact, when we think about it, the motivation we associate with money (once our basic needs have been met) comes from the value it has for us in terms of status, financial autonomy, achievement... A value that cannot be quantified in purely utilitarian terms because it is relative to other values (time spent with family, leisure, recognition, honesty, etc.).

Key Learning Points for the Safety Leader

Along with beliefs, values are the second pillar on which our attitudes are built. Our values are the *affective dimension* of our attitudes, and as such, they exert an even more direct influence than our beliefs when it comes to guiding our perception, our attention, our choices, and our behaviour.

Our values, defined simply as what matters to us, are always relative. The value we give to health and safety, the importance we give to the prevention of future suffering (our own and that of others) compared to the importance we attach to the constraints we would have to submit to now to safeguard them, completes our attitudes.

How to overcome the biases that make us favour the present over the future? Beyond the limited effects of communication campaigns with an affective component, it is the leaders who must fully commit themselves to demonstrating the value that they and the organization place on safety, so that their teams can perceive it and become imbued with it. We will explain the mechanisms at play and the approach to follow in the final chapter.

TOOLBOX – Tool No. 7: Walk the Talk

Every parent knows that it is by demonstrating their daily attachment to certain gestures (like putting away their shoes when they get home), certain expressions (good morning, please, thank you, etc.), and certain rituals (washing their hands before eating) that they pass them on to their children (Figure 61). Inspired speeches and/or irate yelling can neither replace nor erase

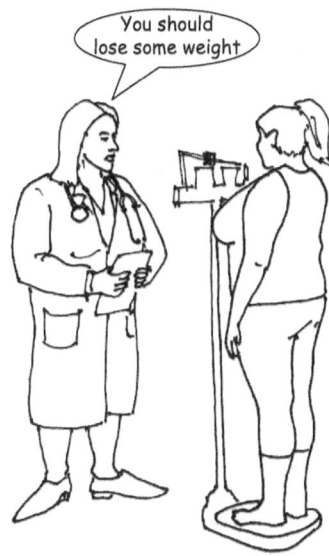

FIGURE 61   Exemplarity (drawing by author).

the messages we send through our actions, which, if they are systematically well-directed and coherent, end up being far-reaching.

Along with immediate and certain/constant/coherent feedback to refocus in the event of deviation and to recognize adequate behaviour (PIC-NIC, presented above), exemplarity is the second fundamental pillar that all leaders must integrate into all their activities to demonstrate the value they assign to something.

It appears as simple and common sense. And yet...

No parent is perfect, and no manager is either. Even the best of us is from time to time prey to mental states that can interfere with the most consolidated of our attitudes, as we shall see. Just as every manager is sometimes confronted with states of affairs that cannot be resolved by will alone, however strong that will may be. Poorly designed equipment, ageing installations, machines that are "not quite up to standard"... these are all situations that could be perceived by our team members as contradicting the value we place on safety, but which we cannot always rectify overnight.

Does this mean that we cannot embody this value? Not at all. We are not condemned to wait until all the material and organizational problems have been resolved before doing so. For that matter, would it even be possible?

It is by making a concrete commitment to significant progress to improve safety in the workplace, and by allocating as many resources as they can, that safety leaders demonstrate the value they place on safety.

We must, of course, remain attentive to the perception that each worker will have of this commitment, as this will always be subjective and inevitably critical. An operator of line A may not be aware of all the efforts that the organization and its management are making on line B, which was in a much worse state of repair. Or maybe they have heard about it, but it does not necessarily impress them because they are more preoccupied with their own worries.

Without indulging in self-congratulation, safety leaders must advertise the progress they are making in their safety-related projects and programmes, and also explain the reasons and imperatives that are delaying those that are not progressing as planned to establish and maintain the credibility of their commitment.

ETHICS

Once it is understood that safety is an attitude based on beliefs and values, the **ethical dimension** of choices and other deliberate behaviours relating to prevention can be addressed appropriately and with the right arguments. We can simply define our ethics as our fundamental ability to make moral judgements on the basis of our *values*. As we have said, these values are cultural, and so are ethics.

According to Kant, "when moral worth is in question, it is not a matter of actions which one sees but of the inner principles which one does not see" (Kant, 1797). For him, *moral virtue* would consist of choosing to act in conformity with the *moral law*, regardless of the consequences for the actor. Thus, he opposed *moral duty* to *inclination* (which we call attitude), which, as we saw, he defined as the preference for some states of the world over others as determined by what one finds *pleasurable*. We will see that the neuroscience of motivation tend to contradict this neat separation.

DID YOU KNOW? START OUT WITH WHAT IS RIGHT

We spoke earlier of how a misguided notion of pragmatism can get in the way of strategic thinking. It can also lead some individuals to dismiss certain values, such as safety, in their quest for "practical" solutions to complicated problems.

In his approach to effective decision-making, Peter Drucker told us that once the problem has been defined, the decision-makers must begin their search for solutions by establishing *what is right*, not in the sense of what is true or correct but in the sense of what is *good*. This involves referring to their hierarchy of values and determining whether a decision involving a certain amount of risk-taking is compatible with the value they place on safety.

Only then they can consider what compromises are acceptable to integrate the imperatives imposed in the light of the competing values that support them (customer service, productivity, profitability, etc.) and find *the right compromise* (Drucker, 1966).

Emphasizing the *value of safety* and making it more important than any other competing value amounts to introducing prevention as a general and overriding duty in what might be called the "code of ethics" of the profession and the organization, applicable to all workers regardless of their role, function, or hierarchical level.

These codes of ethics, this "safety code of conduct", are formidable tools for regulating behaviour, especially in contexts characterized by VUCA (volatility, uncertainty, complexity, ambiguity), for managing intractable systems that do not lend themselves to effective regulation, and more generally for any work situation where there can be no permanent supervision.

Even in systems that are more or less tractable and perfectly supervised, ethical regulation is present and necessary: firstly, as a dogmatic framework for establishing the rules, and secondly for making the necessary trade-offs. Because standards are never perfectly applicable in 100% of cases: a given behaviour may comply with the rules but be dangerous, or conversely, it may be prohibited but be safe, or even provide greater safety than the one prescribed.

Breaking a rule makes the behaviour *punishable*. But with or without a rule, behaviour that goes against a value is always *reprehensible*. Making safety a key part of our values and ethics provides us with a double regulation that acts and consolidates in a double positive feedback loop: self-regulation, complemented and reinforced by regulation through the eyes of others. We will pursue this argument in the next section on motivation.

But let us first focus on discipline and penalties: this involves characterizing the different types of "unsafe acts", distinguishing between violation and error. James Reason proposed an initial classification which he illustrated as follows (Reason, 1990) (Figure 62).

We have studied different types of mechanisms, whether physiological, attentional or memory-based, which operate at a subconscious level and can be the source of unsafe acts. Given their involuntary nature, they cannot constitute a violation.

However, not all intentional behaviour can be considered blameworthy. Among these deliberate acts, a distinction must be made between those that Reason called "thinking errors" and those that he called "violations".

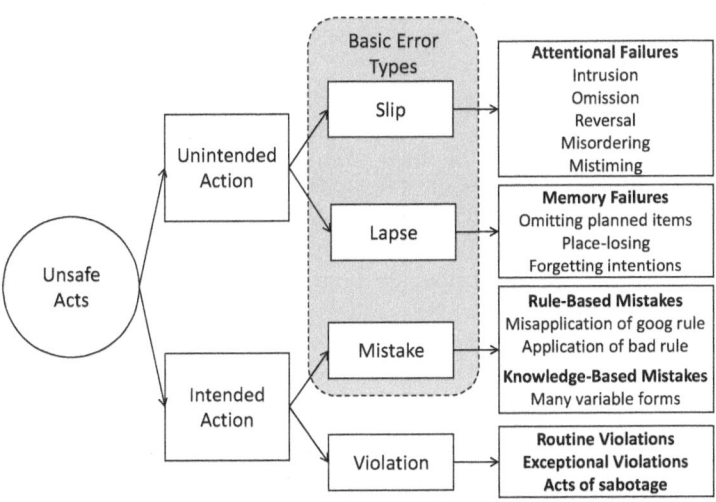

FIGURE 62   Classification of dangerous acts. (Adapted from Reason (1990))

FOCUS ON... THE SUBSTITUTION TEST

Reason later refined his model by incorporating a grading of "culpability" and, above all, what he called the **substitution test** to distinguish "excusable error" from imprudence, negligence, and fault (Reason, 1997). This test consists of asking ourselves whether someone else, with the same skills and subject to the same constraints, would have acted in the same way in an equivalent situation (Figure 63).

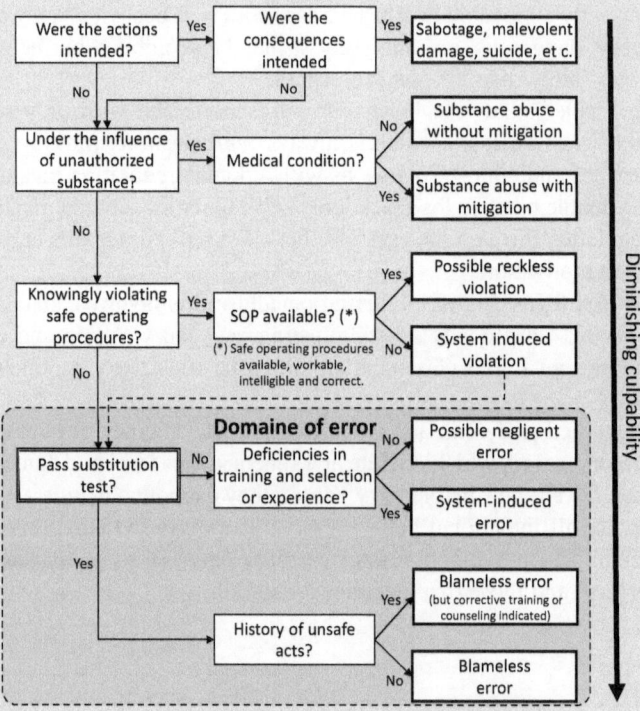

FIGURE 63   The substitution test. (Adapted from Reason (1997))

This way of characterizing dangerous acts remains operational and does not enter into legal considerations. It should be noted, however, that laws and regulations in many countries have adopted the same principle. For instance, article 121–3 of the French Penal Code punishes imprudence, negligence, and breaches of a duty of care "unless the perpetrator of the acts takes the normal care, given his functions, his skills, and the power and means at his disposal", thus following the same approach.

In light of what we have seen in the previous sections, we can consider that not only involuntary acts are "excusable" but also all those resulting from logical thinking based on inappropriate paradigms. This does not mean that they should be ignored: they must give rise to immediate and adequate feedback as per the PIC-NIC technique presented above.

However, those who go against the value that the organization places on safety and therefore its "code of ethics", even if there is no breach of a rule, will have to be seriously reprimanded and eventually lead to the necessary sanctions.

Key Learning Points for the Safety Leader

In the absence of a specific command or safety rule imposing or prohibiting a particular behaviour (which is very often the case), regulation and eventual penalties can only be based on the "ethics" of the profession and the organization insofar as they incorporate the general and essential duty of prevention.

This deontology, or code of ethics, must itself be built on the value that the organization places on safety and on the beliefs that go with it.

TOOLBOX – Tool No. 8: Discipline in Support of Values

After all the information that we have presented about perception, concentration, memory, and rationality – and their inherent limitations – we have to admit that we can all make mistakes. This understanding should give us a dose of benevolence when it comes to managing deviations, but that does not mean we should fall into blissful angelism.

As we stated with the PIC-NIC method, unsafe behaviour must be immediately and systematically addressed. This is a cornerstone of both behavioural safety and the values-based approach to safety. Both schools theorize that the feedback we provide, when it is done properly, tends to reduce the frequency of the behaviour targeted without necessarily resorting to penalties, and this is borne out in practice.

But it would be naive and unrealistic to think that it works all the time and with everyone. Because of their personal and/or professional history, some individuals may have beliefs and values that are too far removed from those underpinning the preventive behaviours we wish to introduce, and may thus resist our best efforts to pull them towards our safety culture.

These cases must be dealt with firmly, because failure to do so not only puts the individual and their colleagues at risk (and creates potential liabilities through our inaction in the event of an accident): this would also weaken us as leaders by undermining the perception that others would have of the importance and therefore the value we assign to safety.

It is vital to address these behaviours consistently, because there is nothing worse when it comes to discipline than giving an impression of *arbitrariness*.

This is why many organizations introduce "golden rules", "life-saving rules", or other "zero tolerance" rules, as well as "non-negotiable" principles and attitudes, with the accompanying range of penalties for any breach. These grids are necessarily generic because, once again, in real life and its intractable systems, we cannot codify everything. But it is important to give ourselves guidelines in this area and to stick to them.

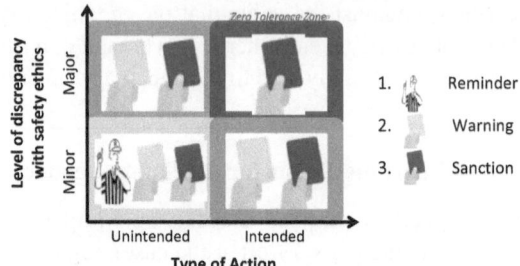

FIGURE 64 "1–2–3" feedback and penalties matrix (drawing by author).

Here is an example of a matrix that is simple to explain in the field (Figure 64). There is no precise reference to the potential severity of the risk, the number of reminders before a warning, or the number of warnings before a sanction, let alone the possible types of sanction. All this is left to the discretion and good judgement of management, under the aegis of the Human Resources department and in compliance with applicable labour codes and regulations.

A word of advice to avoid disparities between the inevitably different rationalities deployed by different managers and therefore reduce "noise" in their judgement: it is a good practice for them to discuss these cases during their management rituals to "benchmark" each other and establish a sort of more or less formal case law.

THE TRIGGERING OF CONSCIOUS ACTION

"In the beginning was emotion".

Louis-Ferdinand Céline

Beyond our automatisms, there seem to be two circuits that lead us to our judgements, choices, and behaviours, whether deliberate or spontaneous: on the one hand, our logic and rationality, based on our paradigms and riddled with shortcuts, biases, and limitations; and on the other hand, our affects and emotions, underpinning our values and indirectly our beliefs, which in turn form the basis of our attitudes and ethics.

These two circuits complement each other and tend to be consistent. But what happens if there is a discrepancy between our logic and our ethics, or between our ethics and the instructions we receive? And more generally, what triggers us to take action?

DID YOU KNOW? ACTION PRECEDES AWARENESS OF ACTION

American neurosurgeon Benjamin Libet made a surprising discovery in 1983: he asked individuals fitted with electrophysiological sensors to flex a finger and to indicate the precise moment at which they decided to flex it. He observed on the electroencephalogram (1) that the individuals were aware of their decision some 350 milliseconds *after* the execution signal sent to the muscles by their brain was picked up by the electromyogram (2).

> The interpretation of these experiences seems to indicate that our actions are triggered at a subconscious level and that it is only afterwards that we become aware of them (Libet, 2004).
>
> 1. Electroencephalogram: examination which records the electrical activity of the brain.
> 2. Electromyogram: examination which records the electrical activity of nerves and muscles.

MOTIVATION

Without getting into epistemological debates, we will simply consider **motivation** as "what drives us, what makes us do things", thus leading to the phenomena we are concerned with in this section: reflection, decision-making, and deliberate observable behaviours relating to safety at work.

The notion of motivation has two sides: the *what* (preferences or the orientation of behaviour) and the *how* (the dynamic aspects or the triggering of action).

In relation to the first dimension, what motivates us, we can develop on what we have already explained about values and go down to the most elementary level to see that, in the final analysis, every individual seeks to increase pleasure and avoid suffering. This postulate of classical hedonism (Moore, 2019) was well demonstrated by B.F. Skinner on rats to whom he gave the means of stimulating their own pleasure-related nerve centres in the hypothalamus, and who preferred to die of hunger rather than stop stimulating themselves.

The same principle applies to our species, with the difference that, because our psyche is certainly more sophisticated than that of rats, we humans have many disparate sources of happiness and well-being that motivate us in our quest to live an "objectively desirable" or *eudaimonic* life. Even Epicurus, who famously maintained that the eudaimonic life is a life of pleasure, admitted that such pleasure also derived from a life of virtue.

However, in a hierarchical and Taylorist working environment impregnated by McGregor's Theory X, we tend to think that it is financial incentives alone that motivate the individual to work and, by extension, to follow injunctions, rules, and safety instructions that impose constraints, solely to avoid sanctions and preserve their livelihood.

> ### DID YOU KNOW? OBEDIENCE TO AUTHORITY: MILGRAM'S EXPERIMENT
>
> In the early 1960s, psychologist Stanley Milgram conducted his classic experiment on obedience at Yale University (Milgram, 1974). Summoning them individually to take part in a fictitious study on learning in return for a small payment, the experimenter asked 40 men aged between 20 and 50, with

different levels of education and occupations, to administer electric shocks of increasing intensity to another individual, supposed to be another volunteer (in fact, he was an accomplice in the experiment, simulating pain because the shocks were not real) if he did not give the right answer to certain questions.

Despite the fact that the shock-generating device warned of the supposedly increasing severity of the shocks, the cries and pleas of the accomplice, and the protests and very obvious signs of conflict among the subjects in the experiment, 100% of them continued the shocks up to 300 V according to the experimenter's orders, and 65% of them obeyed these orders to the end, reaching the maximum of 450 V (which would have been fatal if the shocks had been real).

This experiment, which has been reproduced in various forms and whose results have always been confirmed, demonstrates the powerful influence that orders from an **authority** figure have on us. Education from an early age, in our family homes, then at school, and finally at work (with a stint in the army for some), has inculcated this underlying **value** in us.

This can be good for safety, as long as line managers give instructions that go in the direction of prevention, but it can also represent a danger if line managers ask their subordinates to take risks!

Another very important point stems from this phenomenon: we can have sometimes insurmountable difficulties with people who are resistant to authority (as a result of psychological imbalances or traumas that often date back to their childhood), a tendency that can even be pathological. These individuals can take risks and put themselves and others in danger with the sole aim of going against what is asked of them.

This so-called **extrinsic motivation**, the stick and the carrot, or "simple" obedience to authority (which in reality calls on psychological levers that are not so simple), has very real effects but is insufficient to effectively influence behaviour in a work context where, let us say it again, supervision and control systems are never perfect and where the individuals always have a certain amount of discretionary power to carry out the tasks for which they are responsible.

This is where our attitudes come into play, as we have presented them. Once our beliefs and values have been acquired, our attitudes towards safety become the **intrinsic motivation** that guides our choices and directs our behaviour.

Ultimately, our attitudes play this role on the basis of the same somatic mechanisms of the pursuit of the positive experience/avoidance of the negative experience that was highlighted by Skinner. As mentioned above, acting against our attitudes is unpleasant, if not impossible, except by direct order or major constraint. Acting against our beliefs puts us in a state of cognitive dissonance that makes us uncomfortable, and acting in a way that runs counter to our values triggers even stronger emotions. On the contrary, acting in accordance with our beliefs reassures us and acting in accordance with our values comforts us.

And it is this notion of **emotion** – which we have already introduced and that we will develop in the next section – that enables us to explain the second component of motivation: its *dynamic dimension*, or "how" the action is actually triggered.

FOCUS ON... NEEDS AND MOTIVATION

Among the many theories on motivation, Maslow's "pyramid" is a classic. In the 1940s, Abraham Maslow proposed his **theory of the hierarchy of needs** – which he never represented in the form of a pyramid – according to which we will tend to focus our efforts on satisfying the most primary needs first.

Once these are reasonably satisfied, other needs can emerge. In this way, we gradually work our way up to the top of the pyramid. According to Maslow, just after our physiological needs (the most pressing) comes our need to feel secure. These first two categories of needs are often referred to as *basic human needs* (Maslow, 1943) (Figure 65).

FIGURE 65   Representation of Maslow's hierarchy of needs (drawing by author).

In the specific context of work, a few years later, Frederick Herzberg proposed the **two-factor theory**, taking up more or less the same needs studied by Maslow, but to explain motivation at work. He postulated that meeting *basic needs* at work (which he called *hygiene factors*: administration and supervision, interpersonal relations, status, pay, working conditions... and safety) prevents dissatisfaction. So safety is *necessary* for employee *retention* and is a *prerequisite* for employee *commitment* (Herzberg et al., 1959).

But the simple fact of satisfying these hygiene factors does not trigger any particular motivation to go beyond the "least effort". This motivation to make discretionary efforts is aimed at satisfying the other, more elevated needs, which he called *motivators* (work's challenge and meaning, achievement, recognition, responsibility, and development), which correspond to the higher levels of Maslow's pyramid and where we recognize the tenets of McGregor's Theory Y.

Key Learning Points for the Safety Leader

Obeying instructions and orders to work safely is an important initial motivating factor, but it is only effective in work environments where there is constant supervision.

Feeling safe is a *basic human need*, which will also lead us to avoid immediate and obvious dangers. But beyond this need, we can also link safety to the factors that motivate workers. When safety is a shared value, it is easy to associate it with these **motivators** – which we could call *values* according to our definition and which can therefore be associated with *emotions*:

- Following the same rituals as your colleagues, respecting the same rules, aligning your own behaviour with accepted practices within the group, and avoiding those that are disapproved of or "frowned upon" reinforces the feeling of **belonging** and connection with your co-workers.
- Acting in line with social expectations and, in return, seeing recognition in the eyes of others, managers and colleagues, boosts your **self-esteem**.
- When the organization's culture truly embraces the values and beliefs associated with safety, adopting preventive behaviour generates emotions of **accomplishment**, and conversely, taking risks produces emotions of shame and guilt that can be more powerful than those resulting from a simple risk/reward reckoning.
- At the mature stage of the safety culture, to which we will return later, preventive behaviour **transcends** the safety of the individual, who turns towards others and looks out for the safety of co-workers and third parties.

This "contribution to others" is a cornerstone in Adlerian psychology, being the basis of feelings of worth and belonging (Adler, 1938). Research on performance has also demonstrated this seeking of meaning and contribution to be one of the major motivating factors of high achievers, not only at the individual level (Bloom, 1985; Duckworth, 2016) but also at the organizational level (Gulati, 2022).

EMOTION

These philosophical concepts, psychological theories, and related experimental research on motivation have recently found support from the discoveries made by neuroscientists in the field of **emotion**. From Hellenistic opposition between *hêdos* (ancient Greek for pleasure) and *areté* (virtue) to Kant's dichotomy between pleasure-seeking *inclinations* and *moral duty,* many apparent contradictions can be resolved by better understanding how emotions work.

Traditionally, the concept of emotion has covered all affective reactions, regardless of their duration or intensity. We will first look into a more restricted field, whose study was pioneered by Paul Ekman, which limits *basic emotions* to mental states characterized by rapid onset, limited duration, and involuntary appearance – we will return to other mental states and their effects on action in the next section (Ekman, 1999).

Damasio, whose seminal work has already been presented, defines this type of emotion as the chemical and neural responses produced by the brain when an *emotionally competent stimulus* (ECS) is represented (captured by the senses or in the form of a mental image) in one of its sensory systems, such as vision (Damasio, 2003). Our repertoire of these ECS increases with our experience, and once listed, they are detected very quickly in our perceptual memory, upstream of the processes of selective attention and therefore subconsciously, provoking *automatic affective responses* (Damasio, 1994).

The amygdala and certain regions of the cortex (including the ventromedial prefrontal cortex and the cingulate cortex) are among the areas identified as emotional triggers, while the hypothalamus, the base of the prefrontal cortex and certain brainstem nuclei are areas linked to the execution of emotions, producing hormones and other chemical messengers such as oxytocin and vasopressin.

FOCUS ON... TWO TYPES OF EMOTIONS

Among the basic emotions studied by Ekman, Paul Griffiths draws a parallel with Damasio and distinguishes between those arising from *affective programmes* and those from intellectual functions, or *higher cognitive emotions* (Griffiths, 1998).

- **Affective programmes,** such as anger, joy, disgust, or fear, are associated with simple cognitive processes that developed from our evolution and do not require conscious analysis of the situation. They are transcultural and trigger immediate endocrine and muscular reactions that could be described as *instinctive* and which are the result of an adaptation that has directly improved our chances of survival, such as fear of hearing the roar of a wild animal. They correspond to Damasio's *primary emotions* and to the *thalamic pathway* described by Joseph LeDoux, which triggers rapid emotional responses without awareness of the inducing object (LeDoux, 2005). We could also make the connection with Maslow's *basic needs* presented in the previous section.
- **Higher cognitive emotions,** such as envy, shame, or guilt, require more developed cognitive abilities and are the result of social learning. They are therefore not universal in terms of either their trigger or the triggered reaction: individuals from different cultures will not necessarily want the same thing and will not necessarily act in the same way when they do want it. They correspond to Damasio's *secondary emotions* and LeDoux's *neocortical pathway*, which triggers slower, more complex responses sparked by the awareness and evaluation of stimuli.

Note that it is these higher (or secondary) cognitive emotions that can be linked directly to our **values** in the sense given to them in this book.

Emotions are characterized by two dimensions:

- Their *valence*, or the degree of pleasure or discomfort (or even suffering) they cause
- Their *intensity* in relation to the degree of urgency.

The rationalist tradition (which is no longer dominant in academic circles but remains so in most organizations) has customarily seen emotions as a hindrance to our ability to analyse and make decisions, which in the organizational context must in principle be fully objective and dispassionate. Nonetheless, a number of researchers and philosophers, from Darwin to Sartre, have long emphasized the crucial role of emotions in our cognitive processes (such as perception, attention, concentration, and memory, as described above) and decision-making.

More recently, other authors have pointed out the emotions' intermediary evaluative functions between cognitive evaluation and action, as well as the adjustment of the latter (Scherer, 1986; Lazarus, 1991) (Figure 66).

It is therefore through emotion that our psyche generates the **tension** necessary to trigger action, stop it and, more generally, control it.

Ultimately, it is the result of the integration of the tension generated by different values (some of them competing, some of them aligned), each underpinned by different beliefs, that triggers appropriate work behaviour or, on the contrary, unsafe behaviour.

Intuitively, we might think that reflection could prevail over emotion and direct our actions. After all, we see young children act on their emotions until they learn to control them through reflection (although this mastery will never be perfect). But this inflection always takes place through emotion: reflection modulates emotion, and it is the resulting emotion that triggers, controls, or stops the action.

The development of our prefrontal cortex and the higher cognitive functions it houses has not given rise to any new mechanisms to provoke the act: it is content to call on those that already existed.

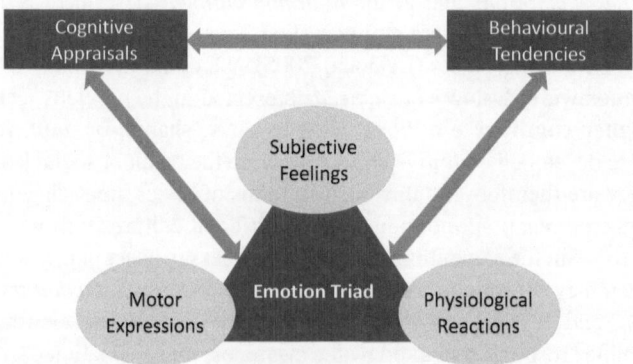

FIGURE 66   Emotional reaction triad. (Adapted from Scherer (2001))

DID YOU KNOW? THE THREE BRAINS: AN OUTDATED THEORY

The three-brain theory was introduced by Paul MacLean in the late 1960s and is still very popular today. According to this theory, our brain evolved in three phases, giving rise to increasingly sophisticated functions:

1. The *reptilian brain* is made up of the brainstem and is responsible for survival instincts (feeding, reproduction, fight or flight, etc.).
2. The *limbic brain*, the central or "subcortical" part of the brain, would then have arrived with the addition of numerous ganglia, responsible for emotions including those of mothering observed in mammals (and which would therefore differentiate them from reptiles).
3. The *neocortex*, the outer layer, which appeared last and is particularly well developed in primates, is home to higher cognitive functions.

However, this theory does not correspond to new data on the neuroanatomy of our brain (or that of reptiles and its descendants – birds – which, like all vertebrates, have structures equivalent to the limbic system and cortex, and some of which display very notable parental behaviours).

All vertebrates have a brain formed around four poles (the brain stem, the cerebellum, the limbic system, and the cortex), the difference being their respective degree of development. In our case, it is the frontal lobe, and more specifically the prefrontal cortex, that is highly developed and supports our working memory and related higher cognitive functions, but it has also acquired a role in managing our emotions, which is therefore not the sole preserve of our limbic system (Dortier, 2011).

FOCUS ON... MR. ELLIOT UNEXPECTED FAILURES

In *Descartes' Error*, Damasio presents the cases of several individuals suffering from neurological lesions and the effects they had on their behaviour, including a patient he calls Elliot (Damasio, 1994).

Previously totally balanced and sensible, a brain tumour caused severe damage to his frontal lobe. He was no longer able to make sensible choices, which devastated his social life, prevented him from keeping a job, and led him to ruin by making financial bets.

However, Elliot had no difficulty in passing intelligence tests and suffered no memory problems. Elliot's problems were **emotional**: for instance, he no longer felt any emotion when shown images of people in pain, despite understanding them.

It was this affective deficiency, which deprived him of empathy and prevented him from feeling (and thus anticipating) emotions such as joy or sadness, that obstructed his reasoning and inhibited his control over his own actions.

DID YOU KNOW? PROCRASTINATION

Procrastination, or our tendency to put something off until tomorrow, can seem "illogical" because it ends up getting us into trouble. But in fact, it follows naturally from what we have just seen: we know we have to do something, but the importance, the value we subconsciously assign to this achievement, fails to generate the emotional tension needed to overcome the inertia of what we are currently doing or to shake us out of our lethargy.

Unfortunately, this is often the case when it comes to preventive action: the notion of a risk that is possible but uncertain, and, in any case, perceived as remote, is not likely to create this tension if the value of safety has not been developed.

Key Learning Points for the Safety Leader

The predominant role of emotion in triggering, controlling, and stopping conscious action is now recognized in the form of two mechanisms:

- Either the stimulus directly produces the emotion which triggers, controls, or stops the action, an emotion which is then rationalized; or
- Reflection, by mobilizing potential mental images linked to affects, generates the emotion that triggers, controls, or stops the action.

But in any case, reflection alone would not be likely to trigger, control, or stop action.

It is therefore through **values** and their affective, emotion-generating influence that we can hope to guide behaviour most effectively.

MENTAL STATES

Different approaches (epistemic, functionalist, consciousness-based, or intention-based) give different definitions of the concept of **mental states** and classify them in different ways. For our practical purposes, we will consider them as affective states, less specific, less intense, and more lasting than the basic emotions that we studied above.

These mental states will influence our behaviour in a less direct and often more subtle way than basic emotions, but some of them can have a very significant impact on our cognitive processes and the choices we make, consciously or subconsciously, when it comes to safety. They are more persistent and will, therefore, affect the individual over a more or less long period of time.

Larry Wilson, founder of a well-known American safety consultancy, highlights the following four mental states as being the most significant:

- **Rushing** can affect the quality of our observation, thinking, and judgements, and it can also lead workers to take dangerous shortcuts in carrying out their activities. The perception of a gap between the time allocated to the task and the time required to complete it is also a source of stress, which, as we have seen, can affect perception, attention, concentration, and memory.
- Physical and mental **fatigue**, which we have already commented on, has a direct impact on safety via physical and cognitive performance. It also has an indirect impact through the emotional state that results from it. Beyond reversible fatigue, from which we recover more or less quickly, chronic fatigue can lead – in addition to the pathological syndromes of burnout and CFS already mentioned – to a mental state of weariness, which can also impair attention, concentration, and motivation.
- **Frustration** can be caused by the same factors as weariness, when taken a step further, but it generally arises when we fail to achieve our goals, whether they are personal, career-related, or task-related. Rightly or wrongly, the individual may project this frustration and look for those around them to blame: internal or external customers, co-workers, management, etc. It can have the same effects as weariness but can also occasionally spill over into anger and lead the individual to sabotage their work, consciously or subconsciously, and put themselves and others in danger.
- **Overconfidence**, which we dealt with previously as a cognitive bias but which can also be approached as a more or less transient mental state in a given context.

Mental states can also become entrenched over time in the individual and become "chronic" through the mechanisms linked to habit formation that we will see in the next section, and even give rise to lasting changes in attitudes if the individual's daily perception of reality changes their beliefs and/or values.

DID YOU KNOW? THEORY OF MIND

Mental states are also similar to attitudes in that they are not directly observable, but this does not mean that they cannot be detected and addressed. The somewhat esoteric term **theory of mind** is used to describe our ability to attribute mental states and attitudes, which are by definition interior mental processes, to ourselves and to others.

We can infer mental states and attitudes on the basis of the behaviours we observe, which enables us to explain them and, to a certain extent, predict subsequent behaviours. Our experience enables us to recognize a whole catalogue of verbal and non-verbal expressions (facial expressions, gestures, postures, etc.) as revealing certain mental states and attitudes. This ability is central to social cognition and indispensable for experiencing empathy and interacting with others by engaging in communication, collaboration, competition, etc. (Duval et al., 2011).

FOCUS ON... FEELING SAFE

At the beginning of this book, we proposed a safety "equation". If we also had to provide an operational definition of this complex and multifactorial concept that is safety, it could be as follows: *a state in which the level of risk is considered acceptable.*

This definition enables us to characterize both the objective state of the situation, in which the danger would be absent or effectively controlled, and the **feeling of safety.** This feeling (or *mental state*) depends as much on the *perception* of danger (discussed previously) as on each individual's *beliefs* about the risk in question (the severity of the feared event, the probability of its occurrence, the level of control exercised, etc.) and the *value* each person places on safety and the other values they associate with it.

Different workers, faced with the same situation, will find it more or less dangerous and will feel more or less safe.

A feeling of safety based on inappropriate perceptions, beliefs, and/or values leads to overconfidence (refer to the example of Captain Smith presented above) and/or carelessness. As we shall see later, some authors even go so far as to list "chronic concern for safety" as one of the characteristics of high-reliability (or resilient) organizations.

Let us just remember that this should remain a recommendation of an organizational nature: we will not dare to demand that every individual should be permanently worried about safety, given the harmful effects of excessive stress (on health but also on safety, via impacts on critical cognitive functions) which we reviewed above.

Key Learning Points for the Safety Leader

As soon as a manager identifies an individual or a group as being affected by a mental state that could lead to unsafe behaviour, she must immediately take an interest in their case. These may be specific and caused by problems outside the workplace, but also be at least partially caused by the organization's socio-technical system, which will be discussed below.

Sometimes organizational or technological changes will provide solutions. Sometimes we need to analyse the individual motivational factors, as we have just seen. And sometimes long-term work on culture will be necessary, as we shall see later.

HABITS

To complete the above, it should be noted that the emotional tension that drives us to act can also arise from simple habit. Studies show that more than 40% of the actions we take are not really the result of decisions or choices on our part (not even spontaneous) but are in fact the result of our habits (Neal et al., 2006).

As we have already seen in the section on procedural automatisms, certain every-day behaviours become so ingrained that they are triggered completely unconsciously. But others, which we will refer to here as **habits**, are carried out *consciously*.

The fundamental mechanisms for the formation of procedural automatisms and these other habits that we consciously follow are the same: the repeated experience of a consequence (a reward or punishment, to use behaviourist jargon) produced by an action or series of actions on our part in response to the appearance of a stimulus. A stimulus that, by repeating the "stimulus–behaviour–consequence" loop, becomes a trigger for action.

In the case of procedural automatisms, a rather specific stimulus becomes the only triggering factor, being sufficient in itself to provoke the action in a given specific context. Whereas in the case of habits, the stimulus is usually less specific and pre-disposes the individual to perform the behaviour and therefore facilitates it, but does not necessarily trigger it (Duhigg, 2012).

FOCUS ON... MR. PAULY UNEXPECTED SUCCESSES

Eugene Pauly had almost died from a very serious case of viral encephali-tis. The virus had destroyed a large part of his medial temporal lobe, and he couldn't remember his age, the day of the week, people's names, or whether or not he had had breakfast. He could not retain any new information for more than a minute or so.

But he could find the way to the toilet without hesitation, even though he could not tell anyone if they asked him. He could not say where the kitchen was, but whenever he was hungry, he went there to get something to nibble on. Not remembering his illness or the injunctions not to leave the house, he would sometimes leave his place: he always managed to get home even though he couldn't draw his own block.

By studying the case of Eugene Pauly, neuroanatomy expert Larry Squire demonstrated that the brain develops lasting habits without using its "regular" long-term memory circuits, but through the basal ganglia.

Most people always park in the same place (in the same section or aisle of the supermarket's or company's car park, if not in the same space). Or always sit at the same table in their lunchroom, if not in the same chair. The stimulus is the fact of arriving at these familiar places, the reward simply being that... so far we have had good results parking or sitting there. Would it be any worse if we switched to another location? We do not know, but since it has worked well so far...

And, above all, we save ourselves the effort of thinking and choosing! Mark Zuckerberg, the billionaire co-founder of Facebook, famously declared in 2014 that he always wore the same model of the grey t-shirt so that he did not have to waste energy on questions such as what he was going to put on (Les Echos, 2014).

In the same way, we tend to arrange our tools in the same place and in the same order to carry out the various tasks in an operation in the same chronological sequence, to stop and take a break in the same place and/or at the same time, etc.

We adopt work routines as well as management routines: the budget review, the Monday morning meeting, the job briefing, the safety toolbox talk...

It is therefore very interesting from a safety point of view to define and establish the most appropriate and safest work habits, as well as the management routines that support prevention. There are many examples of such routines, ranging from stretching and other muscle warm-up exercises to management reviews of the safety system. In the Toolbox we propose one of each, due to their somewhat innovative nature: dynamic risk assessment (DRA) and post-event debriefings (see Tool No. 9).

Just as importantly, these routines also contribute to the development of a safety culture, as *cultural artefacts*, a concept that we will develop later.

But we also need to remember that habits can work against safety, because "bad habits" can develop just as easily. Unsafe behaviour that might be "rewarded" by a gain in time or comfort, for example can quickly become habitual in the absence of effective countermeasures (supervision but also, as we shall see, a culture that promotes the preventive attitudes we have mentioned). This applies to front-line workers, managers, and executives.

Sometimes the effects of habit are more indirect and therefore more insidious: mere exposure to a familiar situation can induce a particular mental state (as described above) likely to affect our performance and behaviour, positively or negatively. For example many drivers who have to contend with traffic jams on their daily commute to work will feel a certain degree of stress as soon as they get behind the wheel, even when they are not under time pressure. They therefore tend to adopt the same aggressive driving style as when they are afraid of being late to work.

Finally, we also have the perverse effect of habits as an obstacle to improvement: routine behaviours that, while not necessarily unsafe, become an obstacle to the deployment of safer practices. Change management techniques, which we will discuss later, will enable us to make these situations evolve.

DID YOU KNOW? ALCOHOLICS ANONYMOUS' TWELVE STEPS

Some habits are particularly difficult to quit, especially those linked to the consumption of addictive substances, which provoke a "craving" involving both physiological and psychological problems during withdrawal. This is the case with alcohol.

A movement started in the United States in the 1930s, Alcoholics Anonymous (AA) has more than 120,000 self-help groups in over 180 countries. It does not claim to replace medicine, but to be its "ally", and it is recognized by the scientific and legal community, particularly in the United States and Canada.

Its method is based on *twelve steps*, a large part of which is based on a belief in a "Power greater than ourselves" (or God, as one may see Him) who can restore us to sanity, to whom we confess our wrongs, humbly ask to remove our deficiencies, and whom we seek to contact through meditation (or prayer) to ask Him to know His will for us.

This very strong spiritual connotation may have prevented AA from further developing in countries with a secular and Cartesian culture, but it can be argued that it is part of the success of their approach. Combining this motivation through beliefs and values, the emotional support of the community, and the "24-hour method" (abstaining from alcohol one day at a time), AA has a remarkable success rate, comparable to that of the most sophisticated pharmacological detoxification treatments.

Key Learning Points for the Safety Leader

Habits are like rails that guide our everyday behaviour, sometimes for the better (good safety habits), but also sometimes for the worse (bad habits that carry risks).

The safety leader must establish and promote good habits and challenge bad ones. Because, as powerful as they are, they can be made to evolve. Sometimes we need a technique, often help (emotional support), but always willpower, which we can link to our motivation and therefore to our **beliefs** and **values** as described above. Here too, approaching safety from this angle provides us with valuable leverage.

TOOLBOX – Tool No. 9: Working and Management Routines

We have just seen that there is a real advantage in terms of mental workload in adopting and following routines that allow us to save cognitive resources: we do not have to ask ourselves questions about the need or desirability of doing or not doing something, we just do it.

Here are two examples, among many others, of good practices in the field of safety that can be ritualized and transformed into habits. The first, DRA, echoes the practice of downtime described above and can become a habit for a group as well as for an individual working alone. The second, the post-event debriefing, can be seen as a collective ritual supported by management.

DYNAMIC RISK ASSESSMENT

Risk assessment is mainly seen as an exercise carried out when designing, installing, and/or commissioning a process or piece of equipment, or when starting up a new activity. This is a lengthy process (of the order of hours or days depending on the complexity of the process or equipment, or the activities to be carried out), often requiring regulatory and bibliographical research, calculations, and tests. It is usually carried out by a multidisciplinary team and is formally documented to ensure that decisions are traceable and to demonstrate due diligence.

DRA is a process carried out in the field by operators (and their direct supervisors if and when they are present), at the point of operation, when they are faced with a situation that could involve aspects that are new or poorly characterized a priori, to identify hazards, assess uncertainty, decide on actions to eliminate or reduce risks, and monitor and review them, in a changing operational environment.

It differs from the verification or control of operational parameters in that, in the case of DRA, there are no instructions or other predefined checklists to guide the assessment process step by step. If there is a procedure or protocol, it is generic, often referring to the doctrine and the main principles of action, with the operator having to choose which to implement in full, in part, or in an adapted manner.

This DRA concept has been implemented, under various names, by emergency and rescue services throughout the world (firefighting, police operations, crisis management, etc.) (ESA ACT, 2022), but it is relevant and can be applied to the world of occupational health and safety as soon as we have teams or isolated operators who may have to deal with scenarios that cannot always be fully foreseen in advance in every detail or that change rapidly as the intervention progresses (installation teams in outdoor environments or maintenance teams, for instance).

We propose here the example of the UK's NFCC (*National Fire Chiefs Council*) based on seven steps: (NFCC, 2018)

1. Assess the situation: identify the hazards and victims; assess the risks on the basis of the severity of the hazards and the probability of occurrence.
2. Decide whether the benefits of the planned intervention are proportionate to the risks. If so, proceed; if not, consider other ways of managing the situation.
3. Choose the protocol(s) to be followed: consider the various alternative courses of action in detail and select the one best suited to the circumstances; assess whether the related means of control can be reasonably implemented and effective.
4. Decide on the action plan and communicate it to all stakeholders.
5. Tactical control: in the case of a team, the person in charge must ensure that each participant has understood their objectives, that everyone's responsibilities are clear, and that everyone has understood the safety instructions.
6. Additional/alternative control measures: as the intervention progresses, new circumstances or information may make it necessary to review the protocol(s) implemented and/or to introduce additional or different control measures from those initially envisaged.
7. Review of risk assessment and effectiveness of controls. In light of this reassessment, the loop may be reiterated.

Once this procedure has been internalized as a routine, interventions are more effective and better controlled, and the risk of error is reduced.

POST-EVENT DEBRIEFING

Just as briefings before starting a job are ritualized in many organizations, debriefings after an event, whether an accident or a near miss, are less so (Blanco-Munoz, 2020).

Substantial amounts of resources are dedicated to investigating and analysing the aftermath of a serious accident, but most organizations do not have the mechanisms in place to collectively learn from day-to-day incidents. This is a shame, since it is these incidents, because of their sheer number, that provide the greatest opportunities for learning and corrective action (Figure 67).

Among the rituals that can be used to collect and exploit this valuable information, we will mention the simplest: the debriefing.

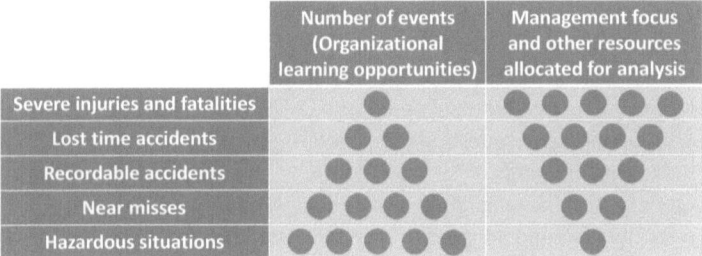

FIGURE 67   Resources allocation paradox (drawing by author).

Debriefing is the process by which individuals who have experienced an event, in whatever capacity, are guided through a purposeful discussion of that experience (Lederman, 1992). This process should provide participants with the opportunity to step back and critically reflect on how the event unfolded and was experienced by each individual and the team.

In a debriefing, the lessons of an event are learned on the spot; learning is direct and based on two assumptions:

1. The experience significantly affected the participants emotionally.
2. Deliberate processing of the experience, in the form of a guided and structured discussion, maximizes its impact as a lesson contributing to the acquisition of a skill or the reinforcement of a mental model. Without reflection, the experience can lose much of its value as a learning opportunity (Pearson and Smith, 1986).

By definition, debriefing is a very open exercise, with little formalized preparation or structure. Here are the few formal elements that typically characterize it:

- *When*: as quickly as possible, typically following a significant event. It can also be a ritual performed systematically at the end of a mission, a simulation exercise, or at the end of a shift.

- *Duration*: from a few minutes to an hour, depending on the complexity of the event, mission, or exercise to be debriefed.
- *Location*: if possible, at the place of the event itself or in a place with direct visual access to its location.
- *Facilitation*: this is generally carried out by the team's management but can also be provided by observers or other external specialists in specific cases, particularly during training exercises and simulations.

The debriefing moderator must guide the discussions, taking into account the following key aspects (Fanning and Gaba, 2007; Manser and Howard, 2000):

- Develop a collegial atmosphere: everyone must have the right and the opportunity to express themselves freely. Care must be taken to avoid accusatory judgements.
- Managers and senior staff speak last to prevent team members from being influenced by their opinions.
- Use open-ended questions to encourage reflection and discussion among participants.
- Openly discuss management strategies and tactics, both the specific ones decided and adopted during the event and the more generic ones drawn up beforehand, but do not discuss those that form part of the organization's doctrine and are considered as operational principles. If any points of this nature seem to pose a problem, they should be noted by the facilitator so that they can be taken up and dealt with by the appropriate body.
- Concentrate the discussion on a limited number of key topics: the succinct format means that discussions have to be restricted to the essential points, which also fosters learning.
- Identify the underlying beliefs that led to misconceptions or errors, and the hierarchy of values that led to inappropriate prioritization of actions and objectives. The previous points are essentially the means of achieving this last, ultimate objective of debriefing, which generally justifies the exercise. The goal is to highlight the discrepancies between the paradigms and the logic applied by the players and those expected by the observer, whether it is a formalized standard, a well-established practice, or the organization's "code of ethics".

SUMMARY OF HUMAN PERFORMANCE AT INDIVIDUAL LEVEL

"Achievement of your happiness is the only moral purpose of your life, and that happiness, not pain or mindless self-indulgence, is the proof of your moral integrity, since it is the proof and the result of your loyalty to the achievement of your values".

Ayn Rand

Throughout these first two chapters devoted to human performance at the level of the individual, we have come across multiple phenomena that are influenced by their beliefs about safety and the value they place on it.

To sum up, in this book, we expose that safety (or prevention, terms which are used interchangeably) is an attitude, based on beliefs and values that the individual acquires in a given cultural context (which we will discuss next). These values and beliefs influence their perception of the environment and direct their attention, are the basis of their affects and emotions, frame their thinking and thus guide their behaviour, both spontaneous and deliberate.

The beliefs to be put forward in each organization are rather specific to its activities and context, but the value of safety (and the other values that may be attached to it, such as family, well-being, etc.) is almost universal. Before moving on to the collective level and seeing how leaders can effectively develop these values and beliefs within their organization's culture, here is a summary outlining how the relative importance that each person gives to safety impacts them on a daily basis (Table 1):

Why is it "important that safety is important" for workers?

Table 1

Studied Phenomenon	Safety is important to me	Safety is not important to me
Predictive coding of perceptual memory	I'm more aware of risks and safety instructions, which I perceive better	Risks and safety warnings are just one of many signals that I can easily miss
Cognitive salience of hazards and safety features	My attention is automatically drawn to risks and safety features	My attention is automatically focused on what interests me and not necessarily on risks and prevention
Spontaneous and deliberate motivation to focus attention on dangerous tasks	I concentrate more on tasks involving risk	Inattentional blindness / habituation
Subconscious inhibition of irrelevant stimuli	I'm less likely to be distracted when carrying out a risky task	I would be distracted from carrying out the risky task if other "interesting" signals appeared
Spontaneous and deliberate motivation to manage fatigue, hydration, nutrition, sleep and lifestyle	I'm more alert, more vigilant, think more clearly and react more quickly	I'm tired, I have trouble concentrating, my thinking is muddled and I lack responsiveness
Spontaneous and deliberate motivation to avoid dividing own attention	I do one thing at a time	I'm doing or thinking about something else while I'm carrying out a task on "autopilot".
Deliberate motivation to anticipate procedural automatisms and reflexes	I take a moment to think before carrying out a risky task, especially if it is routine	I don't ask myself questions before doing a routine task involving risks

(Continued)

(Continued)

Studied Phenomenon	Safety is important to me	Safety is not important to me
Spontaneous and deliberate motivation to store safety-related information in long-term memory	I remember information related to risks and safety	I quickly forget risk- and safety-related information
Availability of accident, risk and safety information in memory	I assess the probability of risks conservatively	I tend to underestimate the probability of risks
Confirmation bias	I am receptive to information that supports prevention and sceptical about information that tends to justify taking risks.	I'm resistant to information that supports prevention and readily accept information that tends to justify risk-taking.
Group effect	I don't hesitate to challenge dangerous situations or unsafe behaviour	I accept the risks that are generally accepted without question
Illusion of understanding	When in doubt, I raise questions and/or adopt a preventive stance by default	I'm satisfied with an incomplete understanding of risks and how to manage them
Paradigms, mental models and thinking	Prevention is central to my thinking	Prevention is not a major factor in my thinking
Emotion and action	I react quickly to risky situations and acting to improve safety is a source of satisfaction	I'm not particularly moved by high-risk situations and taking action to improve safety doesn't motivate me
Habits	I may be brought to change my habits if necessary to make work safer	I'm not going to change my "unsafe habits" unless I'm forced to

4 The Group
Organization, Culture, and Leadership

"Almost everything that characterises humanity can be summed up in the word culture".

François Jacob

The vast majority of professional activities take place within organizations, whether permanent or ephemeral, involving between two and several thousand workers. To address safety at an organizational level, it was imperative to start with the factors we have presented so far, which affect the individual worker (Figure 68).

But in the other direction too, it is just as essential to understand the collective dimension and the phenomena that emerge within it to fully understand the behaviour of individuals belonging to this collective: we have already given some indications along these lines in previous chapters, and we will go into more detail in this one.

FIGURE 68 The individual within the group (drawing by author).

DOI: 10.1201/9781003617341-4

This vision of prevention at a collective level is absolutely essential. It is clear that, if we cannot hope to fundamentally change human nature and "correct" all the limitations we have explored above, it is at the organizational level that we need to consider prevention.

But not just by implementing management systems from an administrative and supervisory perspective: also, and above all, with a view to *leadership*. Because, without underestimating the importance of the former, it is now recognized that steering the behaviour of individuals in an organization over the long-term requires an understanding of their attitudes in their cultural dimension.

STRUCTURE AND RULES

"The purpose of an organization is to enable ordinary human beings to do extraordinary things".

Peter Drucker

Talking about organizations in the broadest sense of the term means lumping together structures that are very different not only in terms of their activity and operation but also in terms of their size. The development of technology, the Taylorism and hyper-specialisation that have accompanied it, and the globalisation of supply chains mean that more and more workers are employed by large companies, which are able to access the capital required to fully benefit from these changes, and many of which carry out their operations on several sites, even in several countries across continents.

At the same time, 61.7 million Americans, totalling 46.4% of private sector employees, worked in 2023 for a small business in the United States (US Small Business Administration, 2023). Similarly, 13.1 million Britons, or 48% of the private sector workforce, worked in 2023 for small businesses in the United Kingdom (UK Dept for Business and Trade, 2023). However, we have spoken of organizations in a generic way up to now, and we will continue to do so in the following pages, because, despite their diversity, there are basically common mechanisms at work.

ORGANIZATIONS

For this reason, we will give ourselves a simple definition capable of covering this heterogeneity and serving our purposes in prevention: by **organization** we mean a group of workers sharing an overall mission, a relational structure, and a set of common basic ways of functioning, all of which contribute to the emergence of a common culture that will in turn support these elements.

FOCUS ON... ORGANIZATION AND WORK

Let us analyse the components of this definition of **organization** and why they are important to our analysis.

Firstly, by **worker** we mean any individual belonging to the organization by the merit of working for it under the aforementioned conditions. Importantly, this not only includes operators but also supervisors and managers up to the highest level.

An organization's **mission** is a coherent set of objectives and goals that it intends to achieve. "Save or perish", the motto inscribed on the badge of the Paris Fire Brigade Regiment in 1942, could not be more explicit about its mission, although it is rare to find such absolute objectives in civilian organizations.

Depending on their sector of activity, organizations will also have more or less laudable, if less peremptory, stated final objectives (design, manufacture, distribute, repair, feed, care for, serve, build, protect, entertain, etc.). Of course, for-profit organizations will always have an underlying and imperative objective of profitability, alongside other intermediate objectives that evolve with tactical orientations: reaching a certain size or market share, developing in a given segment, moving upmarket or reducing costs, etc.

Prevention and workers' health and safety are rarely an explicit part of an organization's mission. When they refer to these subjects in terms of objectives or commitments, they are intermediate objectives, in the sense that their success in these aspects of their operations enables them to achieve their final objectives.

But understanding the organization's mission and ultimate objectives, including those that it does not broadcast, enables us to begin to identify how it views health and safety in its world of constraints and opportunities, and therefore the importance, or value, it places on prevention and the beliefs that underpin it. Because it is at the level of values and beliefs, and therefore culture, that prevention, health, and safety have their place, as we shall explore.

Then, if the organization is to fulfil its mission and achieve its objectives, each worker must be able to interface effectively, in the frame of a **structure,** with all those with whom they need to collaborate. This need to work together, to add and combine strengths to achieve objectives, is the very foundation of any organization, which, as Peter Drucker said, must make effective use of workers' skills and render their weaknesses insignificant (Drucker, 1974).

In addition to the social and affective relationships that develop in any group of people, at work there are operational, functional, and hierarchical relationships. It is the mission that determines the structure, which can range from a perfectly graded and compartmentalised organization chart (as in the case of the fire brigade mentioned above and other quasi-military organizations) to totally informal and more or less "flat" formats.

Regarding the **ways of functioning**, the degree of codification of processes and operating procedures will also vary according to the mission and the VUCA (volatile, uncertain, complex, ambiguous) nature of the activities, their operating environment (from tractable to intractable systems) and their level of risk, which will determine the point of balance between regulated and managed safety that we have already discussed.

These structures and processes, which determine the relative degree of authority and autonomy of each worker, as well as the doctrine, rules, and prescriptions that apply, establish the normative framework for the rights and obligations of everyone within the organization and in carrying out its activities.

As has been widely noted, the structure and processes of organizations carrying out activities involving significant risk tend to be highly formalized (tightly coupled). "Best practice" is codified, based on the organization's own experience as well as that of the industry or profession in question. Workers are informed and trained, and their compliance is ensured by command, supervision, and reporting systems that are generally fairly sophisticated, based on hierarchical structures that clearly define levels of authority.

Ultimately, it is the integration of these three elements – mission, relational structure, and basic ways of functioning – that shapes the **culture** in which all workers (whether operators, managers, or executives) will evolve, and therefore the ethical framework we discussed earlier, based on beliefs and values, which is implicitly imposed on workers by virtue of their membership to the organization.

We will therefore see that in every organization, it is not only the strictly "managerial" parameters that have an impact on prevention. They are very important, and the reader is no doubt familiar with them, but if you thought you could solve all your organization's safety challenges with a purely rule-based, "command and control" approach to safety, you certainly would not be reading these lines.

Indeed, many if not most organizations can be accurately described as "organized anarchies" as James G. March defined them, where conflicts are resolved through bargaining: although they try and provide workers with procedures explaining to them what they must do, in situations of ambiguity and when faced with problematic preferences, decision-making processes derive from the realm of reality, intentionality, and causality to judgements of *meaning* (March and Olsen, 1972). And meaning can only be shared at a cultural level.

We will therefore dispense with the description of occupational health and safety "management systems" and move straight on to the presentation of the cultural approach to safety. But first, in the Toolbox (Tool No. 10), we will provide some advice, informed by the human factors approach, on how to create (and manage) your safety rules to avoid "cluttering" the system.

TOOLBOX – Tool No. 10: Rules for Making Rules

We will not go back over the limits of rule-based safety again. But this approach is nevertheless necessary and complementary to that of the safety culture in a prevention strategy that must be well thought out and balanced.

Every organization establishes rules that become cultural artefacts, as we will see a bit later. They include safety rules as soon as its operations imply coping with risks. These can be seen as the "ground floor" of the safety building, which must obviously be built on the foundations of a sound risk assessment, leaving room for a basement where we will find technological means of prevention to minimise the need for rules (substitution, prevention at the source, etc., in accordance with the hierarchy of control means). Note: to establish and maintain these two underlying levels we also need rules...

Thus, before moving on to more sophisticated subjects, it might be a good idea to ask ourselves whether our safety rules are adequate. This may seem pointless to relatively mature organizations, but experience shows that, without a certain amount of rigour in the administration of safety rules, they can end up doing more harm than good.

So here are six totally empirical parameters that need to be integrated into the proper administration of safety rules (Blanco-Munoz, 2016).

1. **Balance:** Most workers will be willing to walk X yards (or metres) to reach the footbridge to cross the ditch, which is Y feet wide and Z feet deep. But as X tends towards infinity, and with Y and Z constant, the temptation to jump increases logarithmically (see Kahneman and Tversky's graph on loss aversion presented previously).

   So the first rule to observe when defining safety rules (and probably the most important one) is that the level of constraint imposed on the worker must be proportional to the level of risk we are seeking to control and the worker must clearly perceive and understand that this is the case.

   Here are three things to consider:

   a. Let us start with the obvious: always assess the possibility of eliminating the risk, reducing it (in terms of frequency of exposure and/or severity), or putting in place technological means to make the rule less constraining (for example, installing a second footbridge) or eliminating it (organising the work to close the ditch quickly as possible). Always refer to the hierarchy of control means and Reason's "Swiss cheese" model. By forcing ourselves to consider alternatives and therefore resorting to regulatory constraint only as a last resort, we increase the level of safety and, what is more, we reduce the worker's reluctance to accept the rules when we cannot do otherwise. This will defuse the legitimate question of "why don't you take care of making it safer, rather than putting the burden of safety on my shoulders?".

   b. Do not push things too far. Imposing a speed limit of 10 mi/h throughout the site, when some roads are truly risk-free boulevards (perfect surface, good visibility, no crossing of vehicles or pedestrians, etc.), is an invitation to disobedience. This makes both management and the occupational health and safety officer imposing such rules look like zealous ayatollahs, people who do not have a good understanding of real work, or who are too busy (or indolent) to take an interest in the exaggerated constraints that they impose on workers.

If the rule is pushed by upper management, front-line supervisors and safety officers who might share the perception that the rule imposes constraints that are disproportionate to the risk will not try and enforce it. The resulting tolerance may in turn be interpreted by workers as indicating that speeding is not a concern, and they will end up not respecting the speed limit even in traffic lanes where there would be good reason to drive slowly. By extension, other rules may follow the same fate.

c. Communicate effectively about the risk in question, so that the workers affected by the constraints you impose on them are fully aware of the dangers that justify them. We are less tempted to take shortcuts (consider our cognitive biases presented previously) when we understand the reason for a rule. This is the first step towards integrating these elements into our beliefs. Briefings and other safety talks will be welcome, but nothing works better than involving workers in job safety analyses and the definition of control means. We shall come back to these participation processes later.

Establishing balanced rules based on these three principles will also help them to be perceived as fair, see below.

2. **Fairness:** There is little chance of getting your operators to wear safety glasses in a workshop if internal or external visitors (or worse, management!) wander around the premises without wearing them. The rules must apply to everyone in the same way, and anyone exposed to a given hazard must follow the relevant rules, even if the exposure time is short and it could therefore be argued that the risk is lower.

As with balance, the worker's perception of the fairness of a rule is very important. Moreover, this approach also has the merit of simplifying the rule and will help with the next parameter: clarity.

3. **Clarity:** If hearing protection is compulsory in one part of the workshop but not in another, and if it can be removed in exceptional circumstances when a particular piece of machinery is not in operation... you are creating confusion. And for a worker who would already find it difficult to accept a rule, not being convinced of the balance between the constraint it imposes and the safety it provides (see Rule no. 1 above), lack of clarity may be the last straw that pushes her towards non-compliance.

Obviously, rules must be clear enough for everyone to under-stand. But they must also be formulated, displayed, and/or otherwise documented in such a way that every worker has easy access to all the essential information. The safety regulations in most countries require safety data sheets for chemical products used by workers to be available to them. But it is unlikely that they will appreciate having to browse through dozens of pages to see what kind of gloves they will need to handle which container.

Synthetic posters and appropriate signage are generally preferable to documents sitting in a cupboard. When written instructions are required, some organizations impose a limited number of pages and visually accessible formats, with a strong emphasis on illustrations and pictures. If you employ staff for whom English is not their mother tongue, this is particularly important.

4. **Stability:** Once the rules are in place, you should avoid going back over them too often. Obviously, if the process changes or if your risk analysis evolves, you will have to review the entire prevention system, including the rules. But sometimes minor changes bring more confusion than they do good.

And it is not just the headache of document management (see point below). Once workers have established their habits, it is always difficult to change them – valid for both front-line workers and management.

5. **Traceability:** Whatever form your documents take, you need to ensure that they are properly integrated into your documentation system, so that any changes to processes, materials, or instructions are effectively transposed to all the media that need to be modified.

And it is not enough to just change the posted datasheet or the page in the binder: at the very least, you need to inform the workers concerned (and even the safety committee and union representatives in the event of a significant change), taking care to obtain their signatures and keep them carefully... What may seem like bureaucracy to some is an important traceability requirement in the event of an accident.

6. **Supervision:** If you do not have the authority, the means and the will to (i) put in place an effective supervisory system to check the degree of compliance with the rule and (ii) a disciplinary scheme to deal with deviations, you should not presume to impose the rule in the first place.

This does not mean that you cannot try to influence workers' practices by producing and sharing good practices, recommendations, or advice. Call them what you like, but not "rules", because they will not be. And by doing so, the "real rules" lose in visibility and credibility.

The need for supervision does not imply constant monitoring, which is generally not possible. But any worker who is expected to follow a rule must be convinced that, at one time or another, non-compliance may be observed and they will be held to account.

At that point, and depending on the deviation, management can use either the PIC-NIC presented in Tool No. 4 or the discipline presented in Tool No. 8.

Last step: find the right balance

So, the perfect rule should be balanced, fair, clear, and stable, and you should have the authority, the means, and the will to enforce it. It will be very difficult to achieve 20/20 on all these parameters. In most cases, you will have to make concessions on one or another.

A perfectly balanced rule, adjusting the level of constraint as closely as possible to the different operations and their stages, the level of experience and dexterity of the operator, etc., would inevitably be complicated and difficult to supervise.

Therefore, the idea is rather to strike a good balance between each of these parameters and avoid failing in any one of them. Figure 69 shows this exercise which, of course, requires no formula or calculation and is completely qualitative and subjective.

Just one of these factors being "in the red zone" (in the example: a rule that you change too often) can compromise the ability of this rule to play the preventive role you expect it to play.

Finally, we recommend that you use these guidelines not only to establish rules but also to review those that are already in place:

- When you notice that the number of deviations from a given rule is increasing.
- As part of a post-accident investigation and the analysis of the causes that goes with it, when you come to the box "the operator did not follow the instructions". Before asking yourself about the human factor

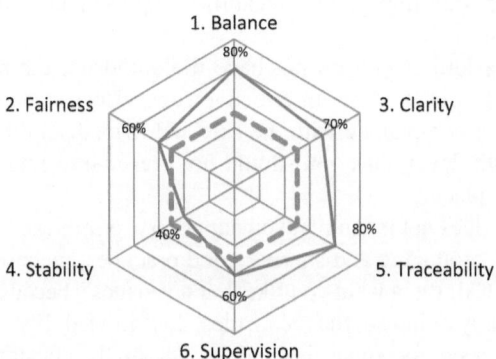

FIGURE 69   Balancing a rule (drawing by author).

in all its complexity, make sure that the rule meets the quality standard presented here.

- Ahead of the launch of an ambitious behavioural safety programme, safety culture or any other approach involving the questioning of worker behaviour, which could lead to backlash if your rules are not adequate to begin with.
- Periodically, as a hygiene measure in relation to your safety manage-ment system. Organizations change, tasks and techniques evolve... and no rule remains valid forever.

WORK COLLECTIVES

Although the cultural approach to organizations can be applied to any type of group of workers that meets the definition given above, it is at the level of close teams, whose members work together very frequently (even if sometimes virtually), that it is easiest to implement. This type of team corresponds to what some authors from the human factors school would call **work collectives**.

FOCUS ON... THE RELATIONSHIP WITH OTHERS IN THE WORK COLLECTIVE

As explained by Christophe Dejours (Dejours, 1995), it is in these **work col-lectives** that the worker is confronted on a daily basis not only with the "*job's reality*", understood as already mentioned as the difficulties against which prescriptions and instructions fail in the execution of the task, but also with *others*: chiefly co-workers and the hierarchy, in conjunction with internal and external customers, etc.

This ego-other interrelationship needs to be fully considered to comple-ment, with a psycho-sociological and cultural perspective, the purely physi-calist or naturalistic understanding that might emanate from the study of the ego-reality axis (Figure 70).

It is the others' eye (not only the customers', the co-workers', the boss's but also the subordinates', etc.) which validates or invalidates, in an intersubjec-tive way, the manner in which the worker adapts the technique to perform the task more efficiently and/or effectively in her eyes, notions which not only

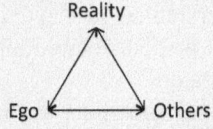

FIGURE 70 Ego–reality–other. (Adapted from Dejours (2022))

implicitly incorporate her beliefs but also all the parameters which are important to her (quality, speed, comfort, economy of effort... and safety) and therefore the hierarchy of her values.

This others' view, therefore, approves or disapproves both the worker's beliefs about techniques and reality and the trade-offs she makes between different values: for example, the precautions and self-preservation gestures she adds at the cost of taking more time or, on the contrary, the shortcuts, deviations, and possible risk-taking at the sacrifice of safety, always based on her local, limited, and multidimensional rationality discussed earlier.

And of course, this others' view is neither objective nor even neutral: it is imbued with the values and beliefs prevalent in the organization.

The closeness of work collectives means that the leader can speak directly, personally, and on a daily basis to each member of the team. In the other direction, the leader can receive significant comments and feedback in the same way, observe difficulties and failures as well as successes, and appreciate attitudes and mental states.

DID YOU KNOW? DUNBAR'S NUMBER

In a famous study published in 1992, the anthropologist Robin Dunbar estimated the number of people with whom an individual can maintain a stable human relationship at 150. He studied the size of the neocortex of different primate species and compared it with the number of associated individuals in their respective groups. By extrapolation, and taking into account the size of our neocortex, the size of human groups should vary between 100 and 230 individuals, with a central value of 150 (Dunbar, 1992).

Beyond this Figure, direct communication does not allow for a sufficient level of trust to maintain stable social relations and efficient operations. To ensure this in larger groups, a more or less elaborated hierarchy needs to be structured and sophisticated rules established.

This proximity brings specific advantages to these work collectives when they are focused on safety (Dejours, 1995; Daniellou et al., 2010):

- There is a space for discussion and dialogue between workers (between peers but also between hierarchical levels) to adapt the working techniques to ensure safety (or at least without compromising it) on the basis of trust, a concept we will deal with later.
- Within the group, abnormal situations are more likely to be perceived, noticed, and reported, and any doubts about the safety of a situation can be discussed quickly between colleagues or with management.

- Both alerts and best practices are rapidly disseminated.
- Work that complies with the collective's "safety ethic" will be recognised directly and immediately, and deviations will be met with equally instantaneous disapproval by others.
- Individual errors can be detected and recovered by another team member.
- New workers' induction/onboarding, integration, and training are made easier, with the transmission of explicit rules as well as non-formalized or tacit knowledge and codes (as already presented).
- The support of others limits the effects of variations in each person's physical and mental state (discussed above) and provides mutual assistance, which reduces stress as well as the individual's effort required to complete the task.

However, certain work collectives' dynamics can play a deleterious role and stand in the way of *cooperation*, thus having a negative impact on prevention:

- Groups in which certain professions or hierarchical levels have an intransigent vision of the activities and risk management practices involved, if there is a discrepancy between this vision and that of the other workers and if there is no room for dialogue between them.
- Groups eroded by technical and/or organizational change, where certain individuals withdraw into themselves and no longer exchange meaningful information with their colleagues, subordinates, and/or superiors.
- More broadly, collectives in which *political games* (which will be discussed in detail below) hamper communication and give rise to conflicts, which may originate in safety-related issues or extend to safety management.

FOCUS ON... COORDINATION AND COOPERATION

Dejours suggests that, at the level of the work collective, we find the same discrepancy that the individual faces between the prescribed job (*work as imagined*) and the actual job (*work as done*). The interactions of each worker with the "jobs' reality" and with others will mean that the collaboration necessary for work in an organization will evolve from *coordination* to *cooperation*.

Just as the worker's practice can either enrich the prescribed task by improving the technique and making it safer, or it can drift towards risk, **cooperation** can also mean either that the collective takes full advantage of the benefits listed above to go beyond coordination and cooperate for safety's sake, or, on the contrary, that its members join forces to endorse or even promote dangerous practices.

Key Learning Points for the Safety Leader

In very compact workgroups, working together on a daily basis and limited in size to around 150 workers, social interactions are fully functional. The other's eyes, and notably the co-workers', validate or invalidate the working techniques adopted by each individual. Cooperation can become a multiplier of safety or, on the contrary, reinforce and perpetuate unsafe behaviours.

It is in such work collectives that leaders have the most direct and decisive direct impact on the group's culture.

Socio-Technical Systems

However, the cultural approach to safety is not just reserved for close or tight-knit work collectives. In fact, it is applicable to all organizations, whatever their scale, as defined above and therefore constituting a **socio-technical system** that may be more or less vast in size, extended or even geographically dispersed, and diverse in terms of cultural substrates (or macro-culture, as we will define them).

FOCUS ON... SOCIO-TECHNICAL SYSTEMS

Frederick Emery and Eric Trist looked at companies in the light of systems' analysis and described them as open systems made up of a *technical-economic system* and a *social system*, the two being interconnected and inseparable. (Emery and Trist, 1965).

The players in these **socio-technical systems** not only develop techniques, tactics, and strategies concerning their work (in the sense we have given above) but also social links, all of which must be coherent if the system is to be stable and able to function efficiently.

It is from the interrelationship of these different components that what we call the organization's culture emerges.

The intensity of these technical-economic and social links between workers will determine the degree of solidity of the culture that emerges within the organization. We can easily imagine how these links are forged between individuals working side by side on a daily basis.

But we also have workers who do their jobs in close physical proximity every day and who do not constitute a socio-technical system – for example, the company's salaried employees and a team of contractors working on the same premises but doing totally independent jobs.

Just as we have multinational companies that have been able to create socio-technical links that may not seem obvious to the casual observer but which act as a backdrop and influence choices and behaviour more or less subtly.

Key Learning Points for the Safety Leader

The professions and the nature of the activities carried out by an organization, as well as its size, with social and functional links that are not always very intense, will make the emergence of a common culture more or less natural and easy.

But as soon as a socio-technical system exists, whether it is compact and informal or vast and hierarchical, it is not only possible but very important to address the cultural dimension of safety, as will be presented in the following sections.

CULTURE

"The culture of a society is the way of life of its members; the collection of ideas and habits which they learn, share and transmit from generation to generation".

Ralph Linton

Once we have characterised the organization, we will define what we mean by its **culture** to study how prevention and safety are integrated and developed within organizations by this means. In keeping with the practical spirit of this book, we are going to avoid complicated philosophical, anthropological, and etymological meanderings and seek an operational definition of the concept, applicable in an organizational context and useful for our prevention purposes.

The simplest formulation, used since the 1980s by Terry Deal and Allan Kennedy, adopted in 1991 by the *Confederation of British Industry*, and still very much in vogue would be "*the way we do things around here*", quite close to the "collective programming of the mind" proposed by Geert Hofstede (Hofstede et al., 2010).

They have the merit of clearly establishing the relationship between the organization and the behaviour of the individuals within it through this "culture". Empirically, we can observe a degree of consistency (if not homogeneity) among the members of an organization when it comes to making technical, organizational, economic, financial, and moral choices, allocating resources and encouraging or discouraging practices. But this definition alone does not give us much clue as to how we can analyse and then direct this cultural phenomenon in such a way as to influence the behaviours that interest us in the direction of prevention.

We will therefore take the UNESCO's definition as a starting point, adapting it slightly to suit the occupational context, and consider culture as the set of distinctive features (material, intellectual, and emotional) that characterise an organization, encompassing its artefacts (including its techniques and practices), as well as its beliefs and values (UNESCO, 1982).

FOCUS ON... CULTURAL ARTEFACTS

To the notions of values and beliefs already presented for studying attitudes, we add that of **artefacts** for studying culture.

Edgar Schein, psychologist, professor emeritus at MIT, and pioneer in the study of organizational culture, used this term to describe all the objects, rituals, and routines that can be observed within an organization, but whose full meaning is difficult to understand for those who have not internalized the organization's culture.

They are therefore objects, tangible or intangible, with a symbolic meaning linked to the values and beliefs shared within the group, which may or may not be linked to any practical usefulness.

The most obvious purely symbolic artefacts include the flags and anthems of nations and clubs, as well as religious symbols and all the ceremonies surrounding them, company logos, uniforms and other dress codes, etc. Objects that combine a practical use with a symbolic nature include professional jargon, tools and gestures specific to a trade, certain meetings and gatherings...

The individual's adoption and use of these artefacts becomes a sign of belonging to the group by visibly demonstrating their adherence to the underlying beliefs and values that they represent.

Their meaning, linked to values and value-linked beliefs, gives these artefacts an affective and emotional load that can be positive (*pleasant,* if the individual adheres to these values/beliefs) or negative (*unpleasant*, if the values/beliefs conveyed by the artefact compete with or contradict those to which the individual adheres).

ATTITUDES AND CULTURE

This definition therefore takes us directly back to the pillars of people's **attitudes**, which we described as being made up of the *beliefs* and *values* they acquire through the *perception* of artefacts, including the practices and techniques deployed around them on a daily basis.

It is impossible to define a culture without recourse to these factors, which are by definition personal. But as in the chicken-and-egg paradox, it can be argued that if individuals share these beliefs, values, and practices, it is because they have acquired them through their live- and working-experience within the said culture, according to Bandura's social learning and social-cognitive theories and his reciprocal determinism model (Bandura, 1986, 1999).

FOCUS ON... BIRTH, EVOLUTION AND DEATH OF CULTURES

This chicken-and-egg paradox illustrates that if we were to look for causal relationships between individual attitudes and collective culture, they could be argued both ways. For we cannot speak of culture until the members of a group share a minimum of these elements, both in terms of the proportion of individuals and the intensity of their adherence; and conversely, it is the actual, day-to-day sharing of these cultural elements that shapes the attitudes of individuals within the group.

In the "normal" circumstances of an organization, i.e. when socio-economic and technical changes are gradual, individual attitudes and group culture will evolve progressively and more or less simultaneously. As no egg has ever magically appeared to give rise to the first poultry, it is also a kind of "cultural selection" process, comparable to natural selection, which will take place in an environment that is not ecological but socio-technical.

The practices and techniques that are tested there and that work effectively will be spread, consolidated, and perpetuated all the faster as the benefits they bring are important. Those that do not work or no longer work will be abandoned, more or less quickly depending on their cost. And the underlying beliefs and values will either be reinforced or eroded accordingly.

The parallel with the evolution of species can also be used to evoke situations of sudden and major socio-economic and/or technical change in this process of organizational culture emergence. Practices and techniques that have historically produced good results can suddenly prove ineffective, established truths can find themselves out of step with the new reality, and what was crucial can become insignificant.

In such upheavals, the organization's culture can either adapt or die. Adaptation involves making substantial changes to artefacts (including practices and techniques) and adjusting underlying beliefs and values to integrate the new paradigm and new priorities while retaining their essence. The death of a culture means the abandonment by individuals of fundamental beliefs that used to be shared, and a radical and disorderly change in the hierarchy of their values. Artefacts that survive purely for their utility are stripped of their symbolic meaning.

The disintegration of an organization's culture means that we can no longer consider it as an organization in terms of our definition: it becomes a "zombie organization", a heterogeneous group without a roadmap or compass. It is bound to fail if its mission, organizational structure, and fundamental ways of functioning are not revitalised with practices that put beliefs and values back in their place, with the necessary adaptations, or completely reinvented on the basis of new beliefs and values.

The same applies to endogenous transformations or internal "mutations" involving profound changes to the mission, relational structure, and ways of functioning that call into question an organization's core beliefs and values (such as the arrival of a new boss with a new vision, takeovers, and major restructuring). The culture will put up resistance in the same way as an organism's immune system fights off an infectious disease or cancer. These mutations will either eventually be defeated or will impose themselves, in which case we will have a new culture and therefore a new organization, even if its logo does not change (although, symbolically, it usually does).

In most cases, the process of developing the place of prevention (as a value in itself, and the beliefs and other values associated with it) within an organization's culture does not fundamentally challenge the organization's other core beliefs and values, and may even not only build on them but reinforce them.

But in some instances, the "mutation" can be significant and trigger major defensive reactions. Later on, we will look at the techniques that leaders can use to help their organization's culture evolve in a way that avoids rejection.

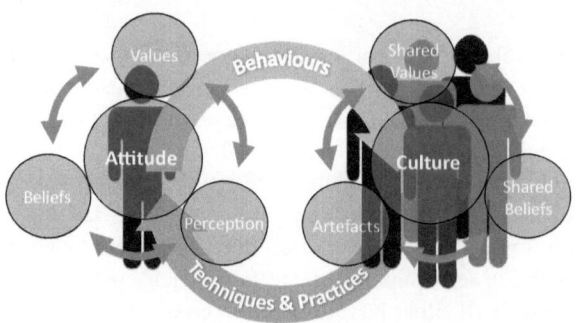

FIGURE 71   The attitudes–culture loop (drawing by author).

We can now combine the diagram we presented earlier on attitudes (Figure 55) with the one presented at the beginning of this chapter (Figure 68) to illustrate this loop (Figure 71):

This model, simplified for the sake of clarity, attempts to illustrate the main concepts presented so far and their reciprocal interactions. On the one hand, we have the individual level (values, beliefs, and perception, interacting with each other and shaping attitude), and on the other hand, the organizational level (shared values, shared beliefs, and artefacts, interacting with each other and shaping culture). Both levels interact in a feedback loop: the individual's attitude, reinforced or challenged by the techniques, practices, and other visible demonstrations of the organization's culture, an attitude that in turn guides the individual's behaviour, which will reinforce or undermine the organization's culture.

In each individual and in each organization, these "bubbles" will be filled with very different elements, elements that will be more or less developed and solidly anchored, more or less coherent with each other, and more or less conducive to safety.

The dynamics that are thus created in different organizations will be very diverse, and even within an organization, the evolution of each individual's attitudes towards a given behaviour will not take place at the same pace. But once a "critical mass" or "tipping point" has been reached, currents set in that pull the majority of individuals along more or less quickly.

Key Learning Points for the Safety Leader

As soon as a group of individuals becomes an organization (with goals and final objectives to be achieved, collaborative and social relationships, and common ways of functioning), the practice of work will tend to align the individuals' attitudes. Being confronted together with the same difficulties and joys, as well as the same failures and successes on a daily basis, makes that a **culture** emerges, carrying with it a conception of safety based on its constituent elements:

- *Beliefs* about what is dangerous and what is not, about the role (function, authorities, autonomy, and room for manoeuvre) of each person, about what is allowed or encouraged and what is not...
- An intrinsic *value* that is attributed to safety, as well as links with other important subjects for the organization and for individuals, which gives it a relative priority, a place in the hierarchy of values of each person and of the group.
- *Artefacts*, including practices and techniques that are more or less safe, formally regulated or not, as well as rituals (meetings, briefings, training, routines, etc.) and physical objects that acquire a symbolic meaning (individual and collective protective equipment, signs and visual codes, indicators, etc.).

As in the case of individual attitudes, these components evolve together in a coherent way, building on each other.

ORGANIZATIONAL CULTURE MODELS

To analyse both the place and interactions of these constituent elements in the culture of an organization and their links with the attitudes of individuals, we will draw on the models of Edgar Schein and Geert Hofstede, which are undoubtedly the most influential and relevant. Both were published in the 1980s and are essentially compatible; they can be used to study the culture of organizations as well as macro-cultures (nations, professions, etc.), which we will look at in the next section. They envisage a culture as a superposition of elements, ranging from the most fundamental to the most incidental.

In Schein's model, we find (Schein, 2017):

1. The **basic underlying assumptions**: self-evident *values* and *beliefs*, fully assumed subconsciously, which determine behaviour, perception, thought, and feeling.
2. **Espoused values and beliefs**: ideals, objectives, aspirations, ideologies, and other rationalisations that individuals *declare* they accept, and which may influence their behaviour to some degree, but do not determine it.
3. **Artefacts**: presented above, these are objects, rituals, and routines that can be observed but whose meaning is difficult to understand for those who have not internalized the culture in question. They are what we perceive and, in turn, produce as cultural symbols.

Figure 72 illustrates this model:

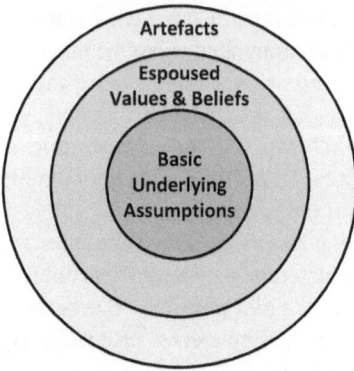

FIGURE 72   Schein's model. (Adapted from Schein (2017))

FOCUS ON... CONFORMISM

Conformism, or the tendency to conform to the accepted norm, goes much further than the group effect biases described above. The classic experiments conducted by Muzafer Şerif (Şerif, 1936) demonstrated that the opinions of others change our own perception of the world, and those by Solomon Asch (Asch, 1987) that we are willing to go as far as to provide answers that we consider to be false to comply with the group's opinion.

Schein's model can be seen as referring to the three levels of conformism described by Herbert Kelman in the 1950s (Kelman, 1958):

- **Internalization**: conformism stems from the unreserved integration of values/beliefs by the individual (full acquisition of the underlying basic assumptions).
- **Identification**: conformism results from the individual's desire to maintain good relations with the group (espousing of beliefs and values).
- **Compliance**: conformism is superficial and seeks above all to avoid exclusion (acceptance of artefacts, including techniques and practices).

In the model proposed by Hofstede, we find (Hofstede et al., 2010):

1. **Values**: general tendencies to prefer certain states of affairs. They operate at a subconscious level and are not directly observable (e.g. preference for normality over abnormality, order over disorder, profits over losses, safety over danger, etc.).
2. **Practices**: observable phenomena derived from the underlying values, which are in turn classified into three levels (from the deepest to the most superficial):
   a. **Rituals**: collective activities whose function is totally or partially unconnected with their practical utility (e.g. protocols for greeting each other, ways of articulating dialogue, certain gatherings).

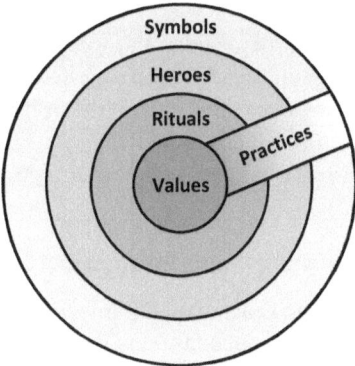

FIGURE 73   Hofstede's model. (Adapted from Hofstede (2010))

  b. **Heroes**: characters, real or imaginary, who embody the ideal values and characteristics of the culture in question and serve as role models (e.g. Ulysses, A. Lincoln, N. Mandela).

  c. **Symbols**: words, gestures, or objects that have a particular meaning within the culture in question (e.g. dress codes, flags, jargon).

Figure 73 illustrates this model.

Although the differences are essentially semantic, in this book we draw more on Schein's model and the jargon he used.

FOCUS ON... THEORY OF BUSINESS

Psychologists and sociologists are not the only ones to have taken an interest in the values and beliefs of organizations. In the 1960s, Peter Drucker, the afore-mentioned professor and management theorist, postulated that to succeed, every organization must define its own theory of the business based on three sets of shared *assumptions* concerning, respectively (Drucker, 1974):

- **Its mission**: the desired results on the basis of shared values.
- **Its environment**: the company and its structure, the market, the customer, and the technology.
- **The competencies** needed to deploy the processes so as to carry out one's mission within one's environment.

Often implicitly, each organization establishes its own theory of business and operates accordingly. For this theory to be valid, it must meet four specifications:

- Beliefs about the environment, mission, and competencies need to be aligned with reality.

- Beliefs in these three areas need to fit together.
- The resulting theory must be known and understood (and we could add *emotionally embraced*) throughout the organization.
- The theory must be constantly tested.

Key Learning Points for the Safety Leader

Values and **beliefs** are at the heart of an organization's **culture**. Assumed to a greater or lesser extent, usually subconsciously, by individuals and forming the basis of their attitudes, they have a direct influence on their behaviour, translating into observable manifestations that shape the collective environment, which provides reference and context.

These observable manifestations, techniques, practices, and other **artefacts** then support and reinforce the values and beliefs within the organization in a self-perpetuating loop.

The way in which safety is integrated into the values and beliefs shared within an organization, crystallizing in its techniques, practices, and other artefacts (and vice versa), will have a decisive impact on workers' behaviour via the affective and cognitive mechanisms described earlier.

MORAL CIRCLES AND MACRO-CULTURES

Now that we have seen how the concept of culture applies to organizations, it is important to understand how these organizational cultures and the influences they have on individuals' attitudes relate to the cultures of other groups to which these same individuals also belong.

Over the course of their lives, all individuals evolve within a multitude of groups, either through day-to-day interaction (family, school, nation, organizations) or through belonging to a "category" (social class, level of education, professional status) that also has its own typical beliefs, values, and artefacts. Successively, but often also simultaneously, we all belong to several of what Hofstede calls **moral circles** (Hofstede et al., 2010).

Individuals find and maintain their place in each of these circles by adopting their values, beliefs, and artefacts, including the practices generally accepted within them. However, the ways in which they do this are not identical throughout their lives.

FOCUS ON... THE DIFFERENT LEVELS OF CULTURAL APPROPRIATION

According to Hofstede, it is during childhood, and particularly before adolescence, that individuals absorb the cultural keys specific to the first moral circles in which they will evolve, the very first being the family. It is during

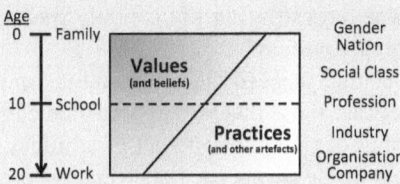

FIGURE 74 Acquisition of values and beliefs (basic underlying assumptions) versus techniques and practices (*artefacts*). (Adapted from Hofstede (2010))

this period that they acquire most of their deeply rooted values and beliefs (or *basic underlying assumptions*), along with the main rituals, symbols, heroes, and other artefacts that accompany and underpin them.

Damasio also considers that it is during this period "that an important repertoire of stimuli coupled with somatic states is built up", even though "the accumulation of somatically marked stimuli only ceases when life ends, and this accumulation can therefore be accurately described as a continuous learning process" (Damasio, 1994).

Nevertheless, as time goes by, it becomes increasingly difficult to alter these neural substrates and therefore this deep layer of our psyche. The left-hand axis of Figure 74 illustrates how, with age, individuals gradually move from an initial period when they are more receptive to *fully acquiring* values and beliefs to a stage of maturity when their integration into new moral circles is rather achieved through the (more superficial) *spousing* of values and beliefs.

From a certain age, individuals will be able to accept the practices of the new groups into which they wish or have to integrate, but it will be difficult to get them to espouse, let alone fully acquire, values and beliefs that are not compatible with those they have already acquired. It is not impossible, but it will take a great deal of *motivation* (and therefore *emotional leverage*, as we saw above) to move in this direction.

This same representation also shows how, in the same way, the cultures of the different groups and categories that the individual gradually integrates are not made up of the same proportions of values and beliefs on the one hand and practices and other artefacts on the other. The axis on the right (not chronologically correlated with the left axis) shows that we share more values and beliefs within the same gender, nation, and social class, i.e. within the moral circles that we integrate during our childhood.

The individual then generally develops within a profession, then an industry, and finally an organization or company, and adopts their respective cultures essentially by accepting the techniques and practices that are promoted and recognised within them.

The influence of an organization's culture on the psyche of its workers is therefore generally more "superficial" than that of **proximate moral circles** (family, club church…) and **macro-cultures** such as nations, religions, or professions. When individuals join an organization as workers, their moral base is already consolidated on the basis of the values and beliefs they acquired during their youth. This is all the more true given that these days working life is rarely spent entirely within a single organization or even a single profession.

The rare organizations that succeed in significantly harmonizing their members' deepest values and beliefs are those that require total dedication from their members and, therefore, a very high degree of alignment of attitudes and behaviours. These organizations invest heavily in advanced indoctrination programmes, with implacable discipline, including the sharing of life-changing, emotionally loaded experiences during long periods of total immersion (e.g. elite military corps).

But it should not be the ambition of safety leaders to seek to radically change the deeply held beliefs and values of workers. As we shall see in the next section, it is above all a question of identifying the underlying beliefs and values that are part of the existing proximate moral circles and macro-cultures and that are conducive to preventive behaviour, to build on them.

Key Learning Points for the Safety Leader

The main difficulty with the cultural approach to safety stems from the fact that workers who join an organization do so with their deep-rooted values and beliefs already in place.

Personnel selection processes in organizations are increasingly focusing on the compatibility, or even alignment, of candidates' values and beliefs with those of the company (which is sometimes called *cultural fit*). Because it is so difficult to assess them in a job interview, many organizations use assessments and other behavioural tests (role-plays, games, etc.) that are supposed to reveal their attitudes in relation to the behaviours expected in the workplace.

However, once they are part of the organization, it is still not only possible but very important to ensure that the values and beliefs linked to safety are developed, as we shall see later.

CULTURAL COMPATIBILITY

Once we have understood that our organizational cultures cannot be decorrelated from their outside world, it would seem appropriate to start by characterizing our cultural environment. So before analysing the organization's current culture, we must begin by studying the macro-cultural framework in which the organization is embedded. This first stage will enable us to draw on the beliefs, values, and artefacts that form part of the underlying macro-cultures (nation, profession, industry) and will give us valuable clues for the subsequent phase.

Indeed, many organizations have tried to implement behavioural safety or safety culture initiatives without first asking themselves this question, and the results have not always been convincing. Using off-the-shelf methodologies, often of Western origin, in countries where some of the expected behaviours are difficult to reconcile with local values, beliefs, and practices, or trying to roll out at a global level programmes designed in the headquarters' country without considering their compatibility with the culture of the receiving countries, have caused many a disillusionment.

We will draw on Hofstede's work to highlight a number of important concepts. By analysing huge databases, he was able to characterise the national cultures of a multitude of countries in relation to six dimensions (Hofstede et al., 2010). These dimensions can be seen as axes with opposing *underlying values and beliefs* at either end.

Among these six axes, there are two that particularly determine the way in which organizations tend to structure themselves and operate, including in the field of safety, by establishing their relationship with authority and rules:

- **Power distance index** (**PDI**), defined as the extent to which individuals not only accept but also expect an unequal distribution of power and authority.
- **Uncertainty avoidance index** (**UAI**), defined as the extent to which individuals not only accept but also expect to be confronted with uncertainty and ambiguity.

Of course, the individuals in the population of each country will vary in these parameters, since not all its citizens have exactly the same distance from power or the same intolerance of uncertainty, but statistically there are significant trends that differ from one country to another.

By putting together these two parameters we obtain the following combinations:

1. **High PDI–High UAI***:* The high PDI is reflected in pyramidal, highly hierarchical organizations, and the high UAI results in a need for systems with very elaborate rules.

    We find in this quadrant most Slavic countries (notably Russia, Serbia, Slovenia, Bulgaria, Poland, Croatia, and the Czech Republic), several Latin-American countries (Guatemala, Uruguay, El Salvador, Chile, Peru, Mexico, Colombia, Brazil, Venezuela, Ecuador, and Panama), many Southern-European and Mediterranean countries (Greece, Portugal, Turkey, Morocco, and other Arab countries, Spain, and France, but also French-speaking Swiss cantons and Belgium), as well as Japan and South Korea.

    Like many other aspects of social relations, safety in these countries is seen from the angle of authority and rules. Those who challenge them (sometimes violently) do so not to abolish them but to replace them with others (more in line with their values/beliefs). The predominant approach is therefore that of rule-based safety previously described.

2. **Low PDI–Low UAI***:* On the opposite side of this two-axis matrix are the United States and most of the English-speaking countries, as well as Northern European countries down to the Netherlands, which have a low PDI and UAI. The organizations are less hierarchical and the rules more generic, often in the form of guidelines. It is the pragmatic analysis of the situation that generally determines in each instance what action to take in a given context and who will take the lead.

    In these countries, prevention is rather understood as a collective responsibility exercised dynamically. They are traditionally more open to the managed safety approach, and it is no coincidence that many of the authors supporting this trend come from these regions.

3. **Low PDI–High UAI**: Germany, Luxembourg, Austria, and Hungary, as well as Israel, are among the countries that combine a low PDI and a high UAI, resulting in organizations where each player needs well-structured regulatory schemes but not necessarily a high concentration of authority or a high level of supervision.

   In these countries, the regulations and systems linked to prevention are rather sophisticated, but the authority for their application comes more from expertise and consensus than from the exercise of hierarchical power.

4. **High PDI–Low UAI**: Finally, China, India, and most of the countries of South-East Asia have a high PDI and a low UAI. This combination favours the emergence of organizations that do not necessarily have highly developed internal regulatory frameworks, with management and control exercised more on the basis of the discretionary power held by those in charge.

   The place of prevention in these organizations depends essentially on the sensitivity of managers to the subject.

FOCUS ON... HOFSTEDE'S 4 OTHER DIMENSIONS

In addition to the PDI and the UAI, Hofstede identified in his early work three other axes for characterizing cultures, to which he added a final one in the latest edition of his reference book:

INDIVIDUALISM VERSUS COLLECTIVISM

The United States, Australia, and the United Kingdom are, in that order, the most individualist countries. Close behind we find Canada, the Netherlands, and New Zealand, followed by most Nordic and Germanic countries and Ireland, as well as France and Belgium.

Employer–employee relations are above all transactional, with neither party generally expecting the emergence of a collective identity within the organization capable of generating a strong sense of belonging, even though some companies have been making efforts in recent years to promote their "employer brand", attractiveness, and commitment.

Organizations that have not succeeded in instilling this "team spirit" throughout their ranks will find it difficult to call on levers such as collective identity, belonging, or pride in the group (even if these can be developed from time to time at the level of teams or work collectives). It should be noted that working on prevention from a cultural point of view can help to foster this collective spirit within the organization, leading to benefits in other areas (commitment, retention, cohesion, etc.).

COMPETITION VERSUS COOPERATION

This axis was termed by Hofstede as *masculinity* vs *femininity* as he understood that some cultural values could be linked to the historical role attributed to men in society (assertiveness, earnings, recognition, advancement, and challenge), whereas other values were more related to the role historically ascribed to women (like modesty, affability, and cooperation). We will refer to this axis as *competition* versus *cooperation*.

Japan, the United States, and other English-speaking countries, as well as Germanic countries, Mexico, and China, are among the most "competitive" countries. On the other side of the spectrum, we have the Nordic countries.

As indicated above, cooperation is generally positive in the field of safety, but when shared values, beliefs, and practices do not promote safety, it can actually increase risks by generalizing unsafe practices.

LONG-TERM VERSUS SHORT-TERM

South Korea, Japan, and China are the top three countries in long-term orientation, followed by several Slavic, Germanic, and Central European countries. The United Kingdom ranks in the middle of the table, and Canada, New Zealand, the United States, Ireland, and Australia follow in that order at the bottom, together with most Nordic, Latin-American, and African countries.

In countries with a long-term orientation, organizations are less likely to obsess about immediate results and are more inclined to look to the future (with the exception of those belonging to English-speaking or Northern European groups, or run by managers trained in management styles specific to these countries, where quarterly, monthly, or even weekly "safety indicators" flourish... we will come back to the subject of indicators later).

This long-term orientation, in which workers attach importance to stability, may make them more receptive to structuring projects and messages that emphasize not only immediate financial performance (and therefore long-term employment) but also health.

INDULGENCE VERSUS RESTRAINT

Latin-American, English-speaking and Northern- and Western-European countries are among the more indulgent (in the sense of tolerating behaviour that deviates from the social norm), while Germany, the countries of Eastern Europe, and Asia show more restraint.

In the latter, the discipline-based approach to prevention is more prevalent than in the former.

TEST NO. 10: Your Organization and Its Macro-Cultures

In light of what has just been presented, we invite you to think about your own organization.

Important: in each question, examine your reasoning carefully and try to avoid falling into the trap of stereotypes, preconceived ideas, and other analytical biases that we saw earlier!

Do you recognize these cultural traits specific to your country in your organization's culture? Or, on the contrary, is it positioned differently in one or more of these areas?

[If you did not find your country in the section above, you can look for it in different websites that offer this information for free, like the one proposed by The Culture Factor Group. Nota bene: neither the author nor his publisher has any affiliation or commercial interests with the authors of this website.]
https://www.theculturefactor.com/country-comparison-tool

If you notice any discrepancies, could they be due to your organization (or a significant proportion of its employees) belonging to other macro-cultures?

- Specific to other nations or geographical areas, for example if a significant proportion of your workers are of that origin?
- Gender-specific, particularly if your workforce is predominantly female or, on the contrary, predominantly male, as is traditionally the case in certain sectors of activity?
- Specific to a particular social class, for example in a traditionally blue-collar employment area?
- Specific to a particular profession with its own very marked codes?
- Specific to a particular trade or industry?

Or are they linked to the cultural particularities of your own organization?

In all cases, carefully ponder the repercussions that the weight of the values and beliefs you have just considered will have on the values and beliefs linked to prevention:

- On which ones could one find leverage and how can you use them?
- Which ones should be tackled and how?

Key Learning Points for the Safety Leader

Not only do the values, beliefs, and artefacts that need to be put forward to develop a culture of safety within an organization need to take full account of the macro-cultural context in which it operates. In reality, in most cases, the development of a safety culture comes down to putting in place and consolidating prevention techniques and practices based on values, beliefs, and artefacts that are already established in the underlying macro-cultures and that would be conducive to safety.

One of the strengths of the cultural approach to prevention is precisely that, very often, it is possible to use certain fundamental beliefs and values, or even artefacts, which are contributing to the preservation of health and safety, as a lever. As we have already seen, the value of human life, attachment to loved ones, and solidarity with colleagues are examples of this type of value. In the other direction, we can also find beliefs, values, and artefacts that stand in the way. As long as leaders do not display behaviors that demonstrate the latter, the development of a culture of prevention is viable.

The models and tools presented in the following sections must therefore always integrate this macro-cultural context and identify the values, beliefs, and artefacts within it that support and those that oppose the implementation of the desired prevention techniques and practices to make use of the former and find ways of limiting the influence of the latter.

Safety Culture

The models described above support our understanding of safety (a term which, it should be remembered, also encompasses health and prevention) as being driven by a series of values and beliefs that underpin the techniques, practices, and other observable artefacts deployed within an organization. We have already cited several authors who have characterized safety, like health and security, as values in themselves (Kahle, 1983; Homer and Kahle, 1988; Schwartz, 2012), values that can be associated with other values held by individuals (family, respect, appreciation, etc.), and with the beliefs to which they are linked.

However, it is probably a misnomer to speak of a "safety culture" (or a "prevention culture", or a "quality culture", a "service culture", or an "innovation culture" for that matter), because no culture is based exclusively on a single family of values, beliefs, or artefacts.

As Reason also notes, a number of characteristics that are not exclusively related to prevention, but are absolutely necessary for it, are also at the heart of so-called "safety" cultures, such as access to relevant information (*informed culture*), transparency (*reporting culture*), fairness (*just culture*), the ability to learn (*learning culture*), and flexibility (*flexible culture*) (Reason, 1997).

Nevertheless, the expression **safety culture** has entered the language of safety practitioners to define the degree to which an organization places prevention-related values/beliefs/artefacts as central and prevailing features of its culture. The most commonly used definition is that given by the International Atomic Energy Agency (IAEA, 1991):

> Assembly of characteristics, attitudes and behaviours in individuals, organizations and institutions which establishes that, as an overriding priority, protection and safety issues receive the attention warranted by their significance.

It is in this sense that the expression safety culture is used, *as a shorthand*, in this book. It is hoped that the reader now has a full understanding of the terms *attitudes, priority, attention,* and *significance* used in this definition.

DID YOU KNOW? 1986, CHALLENGER AND CHERNOBYL

If we had to give a birth year to the expression "safety culture", it would be 1986. Two particularly striking and illustrative accidents took place that year, the analysis of which led to the formalization of this concept.

On 28 January, NASA's Space Shuttle **Challenger** disintegrated 73 seconds after lift-off from Cape Canaveral, killing all seven people on board. It was the first fatal in-flight accident in the American space programme and a shock for the entire country, which watched the event live on television (particularly in schools, as one of the crew members was a secondary school teacher chosen from 11,500 applicants to be the first "space citizen").

The direct cause was the failure of the O-rings on the right solid rocket booster, which had certainly been stiffened by the unusually cold weather in Florida on the eve of the launch, conditions for which the O-rings had not been designed or tested.

During a teleconference on the eve of the launch, in the light of weather forecasts predicting temperatures of −8°C (17.6°F) during the night and −3°C (26.6°F) at the time of the launch scheduled for 9.38 am, engineers from Morton Thiokol, the manufacturer of the solid rocket boosters, expressed concerns about the O-rings. Analyses carried out on these joints on thrusters from previous launches had already revealed their erosion on several occasions, even when not exposed to such low temperatures. Three Morton Thiokol employees and managers recommended that the launch should not take place until the air temperature had reached 12°C (53.6°F).

A break was requested by the manufacturer's management to have a discussion offline, after which they had changed their minds and recommended going ahead with the launch as planned. The NASA representative at this teleconference did not report these exchanges to the launch director. Lift-off took place at 11.38 am, 2 hours later than planned, to give the ice that had formed on the rocket during the night a little more time to melt. The air temperature at the time was 2°C (35.6°F).

On the night of 25–26 April of the same year, reactor No. 4 at the V.I. Lenin power plant in **Chernobyl** – 130 km (80 mi) north of Kyiv in Ukraine, at the time part of the USSR – exploded during a safety test scheduled as part of a maintenance shutdown. Commissioned 3 years earlier, it was the most recent of the plant's four operational Soviet-designed RBMK-1000 reactors, with a capacity of 1,000 MWe (3,200 MWth).

This was the fourth test aimed at validating the capacity of the residual energy from the steam produced by the reactor to power the pumps that were to supply the 28 tonnes of water per hour needed to cool the reactor in the event of a power failure, during the 90 seconds required to start the emergency diesel generators and ensure that they supplied the 5.5 MW necessary for pumping. In principle, this test was essential before the reactors could be put into operation, and the failure of the three previous tests meant that the design of these reactors was out of kilter.

The test protocol, the authors of which were unfamiliar with the details of the plant's operation, envisaged it as a simple electrical test of the turbo-alternators and therefore did not provide for the reactor to be shut down. A succession of delays, unforeseen events, and operating errors, made possible by (and combined with) deficiencies in the plant's design, led to the reactor being "poisoned" by a build-up of xenon-135, and then to it going haywire.

The final attempt at an emergency shutdown by lowering the control rods made of boron (a neutron-absorbing material), which remained stuck at mid-travel in their heat-deformed guide channels, only accelerated the reaction (their lower ends were made of graphite, a material that increases reactivity). The power peak exceeded the reactor's rated power by 100 times, generating an explosive mixture of hydrogen and oxygen by radiolysis of the water. The explosion threw the 2,000-tonne concrete slab covering the reactor into the air, which then fell back onto the reactor and fractured it.

The official and independent investigations into these two terrible tragedies revealed major *systemic failings*: the more or less explicit influence of production imperatives as a priority *value*; *beliefs* that turned out to be false but an inability to question them; a lack of organizational transparency, and pressure to accept the erosion of safety margins at several levels in organizations that were nevertheless responsible for intrinsically dangerous and highly regulated activities, and highlighted the *cultural dimension* of the malfunctions.

We can build on the notions and examples of values and beliefs already provided in the previous chapter and relate them to the cultural framework of organizations, using Table 2 suggested by Patrick Hudson (Hudson, 2001) .

Obviously, these are just examples: the values, beliefs, practices, and other artefacts that each organization develops (or plans to develop) will depend on those of its leaders, as well as on its macro-cultural context, as we have seen.

Key Learning Points for the Safety Leader

An organization may very well be made up of various business lines, teams, workgroups, and other sub-sets, each with its own sub-culture that may differ from the culture of the organization as a whole, while retaining the common base of the latter (otherwise it cannot be said that this sub-set belongs to the organization from a cultural point of view).

However, we cannot accept that a single organization may have different cultures, in particular a safety culture that is independent of the overall culture and/or other cultures (of profit, quality, service, excellence, etc.).

The expression **"safety culture"** (or "prevention culture") is simply a shorthand for labelling an organizational culture in which safety has a prominent place among shared values and beliefs, underpinning widespread techniques and practices that promote safety, and individual and collective behaviour that is generally aligned with it.

TABLE 2
Safety Culture Defined in Terms of Organizational Components

Cultural Element	Definition	Cultural Element Linked to Safety	Examples
Cultural value	What the organization considers important or even sacrosanct	Values linked to safety	The organization considers safety to be sacrosanct and a condition for operating
Cultural belief	What the organization believes about the world, how the world will react to its actions, and what the outside world considers important. Beliefs about what works and what does not.	Beliefs linked to safety	The organization believes that safety is a competitive advantage, that individuals are not the only cause of accidents, that every day that passes brings it closer to the next accident, etc.
Common working practices	Common ways of doing work, e.g. small meetings, piles of memos, and day-to-day project management	Common working practices to guarantee safety	Safety is integrated into design and operations, is the No. 1 item in meetings right up to board level, there is a chronic concern for safety, etc.
Common practices for problem solving	How different types of problem are managed, e.g. project, groups, consultants, panic	Common risk management practices	Risk assessment, cost–benefit analysis, root cause analysis of incidents, proactive research to anticipate incidents, etc.

Adapted from Hudson (2001).

EVOLUTION OF SAFETY WITHIN THE ORGANIZATIONAL CULTURE

"Values provide perspective in the best of times and the worst".

Charles Garfield

Once the concept of culture and its application to occupational safety within an organization have been established, and before looking in detail at the steps that leaders can take to orient workers' behaviour in the direction of prevention, we will focus here on two issues that are relevant to this change.

Firstly, the question of measurement, which is always difficult in the field of prevention and particularly complicated when it comes to its cultural dimension. Then there are the drifts, which mean that the progress made is never fully acquired.

THE PROBLEM OF METRICS

One of the dogmas of management science and all the schools of continuous improvement is that you cannot manage what you cannot measure. It seems to make a lot of sense, and we do share this quest for objectivity... up to a point. We certainly need a factual basis on which to define strategies and programmes, allocate resources, monitor actions, and so on (Blanco-Munoz, 2016).

FIGURE 75 Measuring safety (drawing by author).

The celebrated PDCA wheel (Plan–Do–Check–Act) made famous by W. Edwards Deming cannot move forward if it is not fed with information that enables the initial state of affairs to be visualized, objectives to be set, needs to be assessed, progress to be monitored, and efforts to be reoriented as needed.

But while this quantification may be more or less obvious when it comes to accounting for production and financial results, it is much less so when we consider safety and prevention (Figure 75).

While measurement and the indicators that may go with it are necessary for managing an organization's safety system, they can also have perverse effects on the organization's culture and end up producing unexpected effects that may even be contrary to those intended.

Performance Indicators

Through their ritual use, the organization's indicators acquire the status of *cultural artefacts* as defined above. In addition to factual information, they also convey:

- *Beliefs* about the reality and veracity of what is being measured and our ability to measure it effectively and objectively
- *Values*, as the simple fact of defining indicators, establishing measurement methods, setting up reporting structures, and communicating the results, confer importance on the thing that is being measured.

The indicators par excellence in all areas of management are **performance indicators**. We measure the output of a process, which we can then compare with a quantitatively defined expected performance. Some of these indicators obtain the status of *key performance indicators* (KPI).

In the vast majority of organizations, the ultimate KPI for measuring occupational health and safety is the *frequency rate* (FR). And the further up the organization chart we go, the more safety is reduced to a simple ratio of the number of accidents (with and/or without lost time and/or minor) per amount of hours worked, sometimes complemented with *severity rates* (SR, accounting for working days lost due to workplace injury).

They provide an a posteriori view of safety over a more or less extended period, with a more or less significant time lag. These are perfect examples of *lagging indicators*. Steering a process solely on the basis of these performance indicators is, as we often hear, like driving a car while only looking in the rear-view mirror.

Beyond this observation, which concerns all performance indicators in all fields, the current consensus among occupational health and safety practitioners is that FR/SR are not a good indicator:

- *Unreliable*, as the image in this ever-narrowing rear-view mirror is often distorted, being subject to different definitions and interpretations of the types of accident and to different practices for counting them, which vary from one country to another, from one site to another, and from one time to another depending on the interests of each player (more details in the section on politics later).
- *Negative*, it does not reflect safety, but the lack of it, and ends up contaminating the whole preventive approach with this negativity.
- *Misleading*, because most people who watch them (including executives who are often well trained in economics) cannot tell the difference between variations due to special causes and those due to "background noise", and therefore get excited about peaks or falls that are not statistically significant (Deming, 1986).
- *Unhelpful*, because even when we have the right data and can detect significant trends, the indicator does not provide any clues as to the causes of the variations, nor does it provide guidelines for action when it turns red.

DID YOU KNOW? SIGNIFICANCE AND VARIABILITY

Statistics is a science in its own right and will not be the subject of a course here. But anyone handling figures and trying to understand how they change over time must at least have some idea of how to interpret their significance and variability.

In terms of **significance**, it makes no sense to bother calculating FR/SR for a population of a few dozen workers. It is in organizations of the order of thousands of FTEs that this indicator begins to be of interest (Kjellén, 2009).

As far as **variability** is concerned, we can once again refer to the teachings of Deming (who, as well as being an electrical engineer and physicist, was also a mathematician), who explained that variations in a parameter within three standard deviations of its mean cannot in themselves be considered as indicative of any anomaly. It is when a value falls outside these control levels that we need to look for *special (or assignable) causes*: has there been a change in operating procedure? The installation of new equipment? etc.

Any measurement that falls within these limits (which can be tightened if you like, to two standard deviations for example) comes under what he called *common causes* (or *background noise*), and attempts at interpretation can be deliberately or unintentionally misleading.

If our FR/SR data are indeed reliable (which in most cases they are not), then it is only over the long term that we will be able to see changes, by analysing their 12-month rolling average over several years, for example.

We will not go into further detail on the limitations of these data as a performance indicator, as they have already been the subject of much comment, going back as far as the 1990s (Haines and Kian, 1991). However, it is important to bear in mind the dangers of placing too much emphasis on this single indicator in an organization once we have understood the value of approaching safety from the angle of its culture.

FOCUS ON... CAN INDICATORS KILL?

Here is an empirical list of six toxic effects on an organization's culture, deriving from FR/SR being given pride of place when it comes to measuring safety in the organization. For pedagogical purposes, they are classified into two categories: direct effects on practices and delayed effects via beliefs and values (Figure 76). **Direct effects on *practices*:**

1. **Reduced transparency**: When FR/SR are the most important KPIs, it is with these figure that we associate the various forms of recognition and reward, including bonuses and other incentives. This creates strong pressure at every level of the organization to constantly reduce this figure, sometimes by any means necessary.

   Because of this, some may be inclined not to report accidents considered to be "minor" or to do everything possible to avoid them being classified as "recordable". This prevents the whole organization from getting a realistic picture of the number of accidents and its evolution, benefiting from experience feedback, implementing corrective actions across the board, and ultimately preventing other, possibly more serious accidents from occurring as a result of the related causes.

2. **Quantity versus severity**: Focusing on quantity can divert attention and other resources away from dealing with risks that have a low probability of occurrence but a high potential severity, the kind of hazards that fly under the radar when we only look at FR but which can have dramatic consequences.

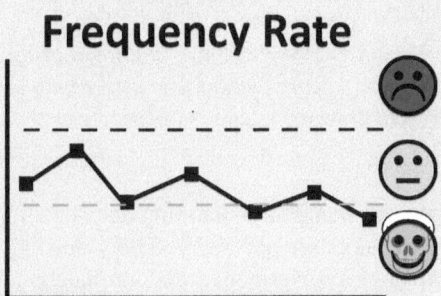

FIGURE 76 FR/SR as main KPIs. FR, frequency rate; SR, severity rates (drawing by author).

3. **Lessened focus**: Let's assume that the practice of systematically discussing safety at management meetings has been instituted. This privileged moment can quite quickly be limited to relaying updates to the FR board. "No accidents last week, very good. Next topic". Talking only about results (good or bad) and simply encouraging teams to "stop having accidents" (or to continue not having any) is not the best way of promoting a safety culture.

Indirect effects via *values* and *beliefs*

1. **Complacency**: When FR/SR begin to stabilize at a reasonably low level (even if this may be due to the efforts of the various players to "massage" the figures as mentioned in point 1 above), a certain complacency can set in at management level, starting with top management, where this indicator is the one that often sums up the safety information they receive (and to which their *perception* in this area of operations is therefore limited). We will come back to this phenomenon in the next section when we talk about *drifts*.
2. **Dehumanization**: By dint of hearing their superiors and colleagues talk about "lowering FR/SR" or "achieving such FR/SR target by such deadline", managers may become desensitized to the human implications of safety. Reducing FR/SR becomes a goal in itself, symbolizing more the political stakes of managerial success than worker safety.

   The link between this figure and caring about their co-workers and their loved ones is then lost. So much so that the first thing some managers think about when they hear of an accident is how it will affect the FR/SR in their area. This is not only humanly sad but also destructive of safety as a value.
3. **Cynicism**: The normal consequence of all the above is that workers in the field end up with the impression that this cold, abstract figure is all that their hierarchy cares about when it comes to their safety. This naturally leads to a certain cynicism when they hear about safety as an organizational value (or culture or priority, depending on the current jargon). Especially if we add to this the clumsy communication on the "cost of accidents" that some safety practitioners, surely with good intentions at heart, tend to do to draw the attention of executives to the subject.

Does this mean that we should stop counting accidents and calculating FR/SR Of course not. These indicators are necessary. Safety must also be understood as a dimension of an organization's performance, and ultimately it is judged by results.

But we must absolutely avoid becoming so obsessed with these figures that we no longer realize what they mean and why we are measuring them. And also understand that it is not this accounting of failures and the analysis of their often stochastic variation that will help us to strengthen our safety culture.

A new understanding of KPIs is beginning to emerge, which reflects this vision: *Keep People Informed, Keep People Interested, and Keep People Inspired.*

Behaviours' Observation

Presented in Toolbox No. 4, data from behaviour observation schemes (in behaviour-based safety approaches) are commonly used to manufacture indicators to monitor changes in behaviour in the field. Observers are appointed, and after receiving training that is often fairly basic and generally does not give them all the keys they need to understand all the dimensions of behaviour, they set about hunting down "risky behaviour" using observation grids, formerly on paper, now more often than not in the form of a mobile application.

With entire teams in some organizations dedicated to this task, it results in colossal quantities of observation sheets, and it is understandable that there is a temptation to establish an indicator on the basis of this easily computable data. But it should be stressed once again that the primary objective of observations in this type of scheme has never been to provide material for producing graphs on changes in unsafe behaviour ratios but to establish a protocol for observation and, above all, for feedback to the people whose behaviour is being observed.

Any attempt to use data from behavioural observation schemes as a leading indicator of safety, prefiguring what will later be measured as FR/SR, must be taken with the utmost caution.

FOCUS ON... DATA FROM BBS OBSERVATIONS

As mentioned when these schemes were presented, there are three main objections to using the data collected during behavioural observations as an indicator of safety:

Firstly, there is (always) the question of **reliability**. Any indicator based on the quantification of observable phenomena can be reliable if the individuals responsible for monitoring them are genuinely committed to transparency. The question is therefore to find a way of motivating them to faithfully report information that may cast their management work in a negative light. Especially when it comes to behaviours that, although they are certainly factual elements, do not necessarily leave material traces (such as an injury or damage). Here we find the same difficulty as in using the number of near misses as a safety indicator: are we even capable of estimating the proportion that we manage to record?

The second question is (once again) that of **significance**. Taking into account the thousands of behaviours of each individual during their working day, and the variability arising from the biases inherent in their observation (such as the accessibility of different activities and workers to observation, or the competence and motivation of observers), we can deduce that the sampling practised by most organizations is largely insufficient to produce statistically significant results.

Finally, there is the question of **correlation** with safety performance, which could eventually justify it being called a leading indicator. Unfortunately, neither the number nor the severity of accidents over the medium and long term seems to be a function of the ratio of unsafe behaviour to the total number of observations made. Knowing that they are being watched has a very significant influence on workers' behaviour at first, but then this influence fades. We therefore find downward and then upward variations in the unsafe behaviours observed, which cannot be correlated with the actual number of accidents over the period, or with the maturity of the safety culture.

DID YOU KNOW? HAWTHORNE EFFECT

A certain form of Heisenberg's uncertainty principle, known as the Hawthorne or **observer effect**, always casts doubt on studies based on observing workers when they know they are being observed.

This phenomenon owes its name to the place where it was first noted: an electrical components factory called Hawthorne Works in the suburbs of Chicago (US). Between 1924 and 1932, a number of experiments were carried out there to measure the impact of various parameters on worker productivity (for the record, Edwards Deming worked there for two summers as part of his doctorate in mathematics during this period).

Initially, changes in light levels in the workshop were studied, then the effects of factors as diverse as the cleanliness of workstations, the presence of obstacles on the floor, changes in the location of certain workstations, and the organization of breaks were analysed.

Productivity increased each time a change was introduced and an observer was present to measure it, but then fell at the end of the study. Several hypotheses were proposed to explain the phenomenon, but the most popular was that put forward by H.A. Landsberger in 1958 on the observer effect, which became known as the Hawthorne effect (Landsberger, 1958).

Similarly, it seems entirely plausible that workers are more attentive and avoid unsafe behaviour when these programmes are launched if they know that an observation sequence is taking place. An effect that may very well fade as they grow used to the drill.

Audits

There are now a number of auditable standards in the field of occupational health and safety management, including ISO 45001, which has already been mentioned. They provide a framework of generic requirements that all organizations must meet to be certifiable after a third-party audit, and can obviously also be used as a reference by all organizations even if they do not intend to be certified.

One of the requirements found in all these standards is precisely to define and implement an internal audit process to ensure periodic verification of compliance with each of the requirements of the standard in question.

This "management system" vision that we have abundantly discussed is very interesting, and the fact of carrying out periodic audits to check that established procedures are being respected is a necessary part of this approach. In the cultural approach to safety, these audits can be seen as *cultural artefacts* (just like the indicators and behaviour observation initiatives mentioned above) whose practical usefulness is coupled with a symbolism that reinforces the importance that the organization gives to prevention.

However, their relevance as a tool for measuring the maturity of an organization's safety culture has not been established. Organizations may have all the formal elements (or artefacts) needed to demonstrate compliance with a standard and yet fail to give them meaning, failing to engage workers in a dynamic where the value of safety and the beliefs that go with it are genuinely shared and therefore failing to motivate the expected behaviours. Conversely, organizations can succeed in putting prevention at the heart of their culture without necessarily adopting a standard formalism.

FOCUS ON... ISO 45001 CERTIFICATION

ISO 45001, which in 2018 replaced the BSI's former OHSAS 18001 as an international standard, is undoubtedly the benchmark standard for occupational health and safety management. Like its equivalents, ISO 9001 for quality and ISO 14001 for the environment, it is based on ISO's High-Level Structure framework and completes the triptych, enabling organizations to integrate their Quality – Health & Safety - Environmental (QHSE) management systems.

One of the major changes compared with its ancestor is the integration of management leadership as the driving force behind the management system, as well as the requirement to consult and involve workers (elements to which we will return later), in an attempt to incorporate not only a certain vision of managed safety, as we discussed earlier, but also of the cultural approach to safety.

It therefore provides an excellent basis on which building your management system with a systematic approach to ensuring that all critical topics (risk assessments, determination of regulatory and other requirements, setting objectives and planning actions, skills management, operational control, performance and compliance assessment, audit and corrective action, management review, etc.) are properly addressed and remain under control.

From a cultural point of view, workers' perception of the rigour put forth by the organization in following these precepts and applying these artefacts can serve to reinforce their belief in management's commitment to safety and to espouse it as a value (third-party certification can also contribute to this last point).

However, if the implementation of this type of system is only a response to an obligation imposed by customers, by shareholders, or by the group to which the organization belongs, if its requirements are experienced as a constraint by the management and/or by the executives, and if this becomes obvious to the workers, the effects on the attitudes of the latter with regard to prevention may, on the contrary, be negative.

We also find in the literature (but also in commerce) (ICSI, 2022) examples of audit grids and other safety culture "diagnostic tools", which have been tested according to more or less scientific protocols – in any case not always available for critical analysis. They have been developed mainly in the oil & gas, healthcare, and aviation sectors, and can be classified into two broad categories.

The first of these categories corresponds to "real" audits: through inspection visits, interviews, and document reviews, the auditors look for evidence to show that the audited organization systematically follows a series of predefined techniques and practices. This is similar to an audit of the safety management system of the organization in question, with a reference framework presenting, in the form of requirements, a way to inquire about what the authors of the grid consider to be the fundamental elements of a safety culture.

The second type of so-called audit is in fact a questionnaire-based survey, which will be discussed in the section right next.

Surveys

To examine their safety culture, a subject that, as we have seen, is in many aspects subjective and intangible, many organizations have turned to the classic tools for studying this kind of thing: surveys. Several vendors offer this type of off-the-shelf tool, some of which allow you to compare your own results with those of an anonymized benchmark of other organizations (HSE, 2022).

These surveys, whether carried out by means of questionnaires completed during an interview (face-to-face or by telephone), in paper format or online, are among the tools of choice for marketing experts for *measuring the attitudes* of individuals in a target population towards a product or behaviour (ranging from consumer products to voting choices) and deducing from them the underlying values/beliefs that it will be necessary to try and influence to steer these behaviours in the desired direction.

Even though research on the subject shows that there is often a limited correlation between the attitudes declared by individuals (typically when they respond to this type of questionnaire) and their behaviour (Fishbein and Ajzen, 1975), a well-conducted survey, based on a well-designed questionnaire, can help to identify, at least roughly, the components of the respondents' attitudes that need to be looked into (values or beliefs) and give us clues as to the artefacts that need to be deployed (or whose deployment needs to be refined) to do so.

There is therefore real value in investing the time and resources to carry out a survey before launching a programme aimed at the cultural dimension of safety, and it is even conceivable to repeat the exercise periodically to assess its success in terms of progress in individual attitudes and therefore in the organization's safety culture.

However, the relevance of these data as an *indicator of safety culture* needs to be nuanced. Several authors signal that such questionnaires are rather revealing of the organization's current **safety climate**, which is more related to the workers' perception of the safety conditions and experiences than to the organization's culture, which is a more complex and stable phenomenon requiring a more in-depth approach to be entirely understood (Tharaldsen et al., 2008). These surveys can be seen as "snapshots" (Flin et al., 2000) that provide a "quick (easy to acquire) and dirty (broad and superficial)" vision of a safety culture (Guldenmund, 2007).

The Qualitative Approach

Some managers may be discouraged at this stage by the difficulty of assessing their "safety culture" and its progress. It is important to accept that it is impossible to perfectly quantify attitudes and culture, just as it is impossible to quantify values and beliefs. But that does not mean that we cannot appreciate and observe these elements in a qualitative way. What many professionals call "feeling" an organization's safety culture.

We have already discussed how the mechanisms linked to the theory of mind enable us to assess mental states, emotions, and attitudes, faculties that are essential for our day-to-day social interactions. This involves face-to-face exchanges with workers, discussions with those who have to make trade-offs, take decisions or follow instructions, and direct observation of what is done and how it is done, i.e. the ethnographic approach to the study of cultures (Dupré and Le-Coze, 2023).

This is not a quantitative approach, but it is nonetheless a factual one, since we can contrast workers' declarations with the behaviour and choices they actually make. It is not easy to produce fancy graphics with this information. But maybe we do not have to put a figure on everything.

DID YOU KNOW? MANAGEMENT'S SEVEN DEADLY DISEASES

We have mentioned W. Edwards Deming several times, his statistical approach to processes and his healthy tendency to rely on data when making management choices. However, it was Deming himself who labelled the fact of "managing by use only of visible figures, with little or no consideration of figures that are unknown or unknowable" as one of the **seven deadly diseases** of Western management style (Deming, 1986).

This may seem to contradict one of his most famous phrases, which was that "without data you are just someone with an opinion", but on closer inspection, this is not the case. This is undoubtedly one of the fundamental differences between management in the traditional sense of the word and the leadership we are seeking to develop: the ability to act, decide, and mobilize teams for the sake of safety on the basis of an opinion, a feeling, or an emotion (a concept which, after what we have seen above, should no longer frighten us).

An opinion built by deliberately reflecting on the emotion resulting from our perception combined with our values and well-informed beliefs (based on a good understanding of occupational risks, human factors, and the cultural mechanisms of organizations) could well be a better guidance than a dubious figure interpreted in a haphazard way.

In this tradition of qualitative evaluation, various authors have arranged the constituent elements of a culture of prevention in different ways to characterize the progressive stages in the evolution of the culture of organizations, according to the presence and prevalence of these elements (Westrum, 1984, 2004; Reason, 1993; Fleming, 2000).

Outside the safety science community, the best-known model is undoubtedly the **Bradley curve**, proposed by Vernon Bradley in 1994 and largely inspired by the ideas published by Stephen Covey a few years earlier (Covey, 1989), but whose rights were nevertheless registered by the former consulting arm of the DuPont company. It theorizes that an organization's accident FR is inversely proportional to the maturity of its safety culture, which can be situated in one of the four successive stages below:

1. **Reactive**: Safety would be based on the survival instinct of individuals, who in principle should just listen to these instincts and above all follow the rules in their own interests. Management does not feel involved and delegates safety to the ad hoc function ("mind your own business").
2. **Dependant**: Management shows a degree of commitment and appoints supervisors to monitor compliance with the rules. Operators receive safety training and are well warned of the disciplinary consequences of non-compliance ("if you don't follow the rules, you'll be fired").
3. **Independent**: the system supports prevention practices. Individuals feel committed to safety by spousing related values that remain focused on themselves ("we take care of ourselves").
4. **Interdependent**: safety is supported by systematic practices, which in turn are based on organizational values and beliefs that give them meaning, such as collective responsibility, transparency, and trust, which prevail in every individual and are proudly displayed as the organization's essential characteristics ("we look after ourselves and others"). This is the principle of *shared vigilance.*

Figure 77 shows the Bradley curve:

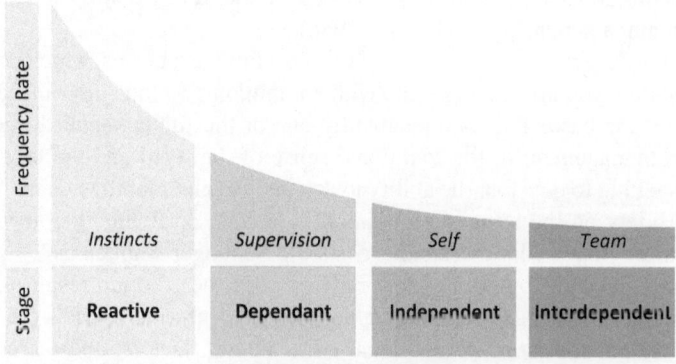

FIGURE 77 Bradley curve. (Adapted from www.consultdds.com/transform-culture/dss-bradley-curve/)

What these scales and other maturity curves have in common is that they provide a qualitative assessment of where a given organization stands in terms of the acquisition and gradual integration at the group level of prevention-oriented values, beliefs, techniques, and practices, until they become central, essential, and permanent features of the said organization.

FOCUS ON... INTERDEPENDENCE AND SHARED VIGILANCE

The notion of **shared vigilance** is far from new or truly original, but it is topical among those who advocate a cultural approach to safety. The general idea is to instil in everyone the feeling that, at work, it is not enough to look after yourself, you also must look after the safety of those around you.

This simple concept is the pinnacle of a mature safety culture and is only fully achieved in organizations that are well advanced in this area. It is the questions, remarks and other daily exchanges between peers on the subject of safety that, by becoming habitual and therefore spontaneous, are ritualized and acquire the status of artefacts, loaded with preventive values and beliefs.

Its development in various countries has perhaps been held back by the suspicion with which anything that might remotely resemble tattling is greeted there (an attitude that is not so negatively connoted in Switzerland, for example, where denouncing an offender is a matter of civic sense – always a question of culture...) or simply "minding other people's business".

In any case, it is certainly not in the spirit of reporting to management that workers are encouraged to be attentive to dangerous situations and more particularly to the unsafe behaviour of others (not only colleagues but also subcontractors and other third parties), nor is it in the spirit of transforming workers into "safety vigilantes".

The objective, which must be perfectly clear and communicated without any ambiguity, is that if you see something that concerns you, you must feel legitimate and safe to go and ask the other person: "why are you doing this?", "what do you think the risks are?", "what do you intend to do to manage them?", "if you agree that the risk is considerable: how could you do otherwise to avoid putting yourself (and/or others) in danger?".

It is the behaviour that should be questioned, never the person. With open-ended questions that show genuine curiosity, on the assumption that this behaviour no doubt appears to her to be entirely reasonable from the point of view of her own local rationality. And with the humility to realize that we probably do not have all the information we need to judge the behaviour; that the other person's reasons may be valid and so may be her strategy/tactics/technique; and that in the end, it may well be us who learn something.

The communication promoting this practice and the underlying mindset must be aimed just as much at the person who notices something as at the person who might be questioned by a co-worker: particular emphasis must be placed on explaining very explicitly that the latter should take these questions in a positive way, in fact as a gift offered to her by someone who perhaps had to overcome a certain feeling of insecurity to come and talk to her, because it is never easy to question the actions of a co-worker. A gift that allows her to broaden her perspective and perhaps realize that the risk she was taking was too important after all.

Kindness and **respect** are the watchwords of this practice, which might seem trivial but unfortunately can be a long way off for organizations where the safety culture is not yet very robust. It is a practice that *emerges* as a result of this culture, and while it can (and should) be encouraged, it cannot be decreed or institutionalized through instructions. Not forgetting that, once again, the political dimension discussed in the next section can represent a major obstacle.

It should be pointed out that, as part of this approach, workers may also be encouraged to note the mental states discussed above, as part of a prevention approach also geared towards *mental health*, thus overlapping with another topical issue, but which may need to be the subject of specific communication, or even special training, given the intricacies of the subject.

Key Learning Points for the Safety Leader

While it is natural to seek to measure safety, and perhaps necessary in the end to a certain degree, safety leaders must understand the limits of this subject as far as quantification is concerned. They will need this understanding, first of all, so as not to waste their time (and that of others) by embarking on the ritual monitoring of indicators that mean nothing and demonstrate the cult of metrics rather than the value of safety, and by trying to weave a story around figures that have no meaning other than that which we impose on them.

Safety leaders will also need this understanding to be able to explain to managers and executives when certain variations in the indicators selected really indicate something and when, on the contrary, they fluctuate for no other reason than chance – see our safety equation presented at the beginning of this book.

Finally, having a good grasp on these phenomena will also be very useful to explain why, sometimes, it is better not to have (or not to highlight) an indicator than to have (or to highlight) a bad indicator, especially if the latter dehumanizes safety. This message, which may seem like sacrilege in the face of certain deep-rooted beliefs held by managers and executives, is likely to be better heard as space is made for other beliefs that are more conducive to prevention, such as the relevance of qualitative assessments.

In the end, these will be no more subjective than the interpretation that might be made of more or less truthful and significant data.

DRIFTS

So it is not easy to measure safety, let alone its integration into an organization's culture. But whether it is quantitatively (FR/SR, unsafe behaviour, near misses... as long as they can be properly accounted for), indirectly (audits, surveys...) or purely qualitatively, we can nevertheless perceive **drifts**: the slow erosion of safe practices and the insidious installation of unsafe ways.

In his work on the *Safety Differently* vision, Sidney Dekker argues that workers "create (the conditions and skills necessary for) safety by practising" (Dekker, 2014). However, if leaders do not make their best efforts to consolidate the place of safety in the organization's culture, continuous practice can also lead workers, on the basis of the local, limited, and multidimensional nature of the rationality that they deploy (as we saw above), to gradually move away from safe techniques and practices.

This happens not only at the individual level but also at the collective level, and above all, it does not just happen at the operator level: it also concerns managers and executives.

So let us take a closer look at these phenomena before moving on to consider how leaders should tackle them.

Practical Drift and Drift to Danger

The phenomenon of drift has been extensively studied in the aeronautical and health-care environments, and we can apply the lessons learned in these fields to almost any work situation. Two complementary approaches describe this dynamic, one based on the concept of **practical drift**, the other on that of **drift to danger**.

Practical Drift

Based on risk assessments and learnings from (sometimes catastrophic) experience, procedures, instructions, and other operating methods are formalized and deployed. The aim of these instructions is to control the system so as to make it as safe as possible, and they tend to be drawn up in such a way as to regulate the work in detail, often from the point of view of a tractable system, as we have presented them.

However, over time, the standards and constraints defined internally or imposed, which were justified by the rationality of the prescriber, seem increasingly disproportionate to the operator carrying out the task. Gradually, the emotional shock of the past accident fades from individual and collective memories, and at the same time, the intrinsic complexity of the system and the conflicting organizational objectives and those of each player become apparent once again.

If safety techniques and practices are not firmly anchored in the socio-technical system by means of the cultural forces mentioned above (values, beliefs, and artefacts, including effective control mechanisms), if the coupling is artificial rather than organic, **practical drift** is inevitable. This concept was introduced by Scott Snook in the 1990s to describe the tendency of techniques and practices to deviate, generally under the influence of economic and operational constraints, from those formally validated (Snook, 1996).

Drift to Danger

The various factors involved and the stages in this process are illustrated by Jens Rasmussen in his **drift to danger** model, which he defines as the "systematic migration of organizational behaviour towards accidents under pressure to optimize costs in an aggressive, competitive environment" (Rasmussen, 1997).

FOCUS ON... THE RANGE OF POSSIBILITIES

Workers, whatever their hierarchical level or function, operate within a range of possibilities delimited by economic, workload, and safety constraints, as illustrated in Figure 78:

FIGURE 78   Space of possibilities (Adapted from Rasmussen (1997)).

Individual workers, and the organization as a whole, adapt their techniques and practices as a function of the pressure exerted by each of these three factors. The competitive environment and the short-term financial logic prevalent in many organizations drive decision-makers and management in a quest for economic optimization, as shown in Figure 79:

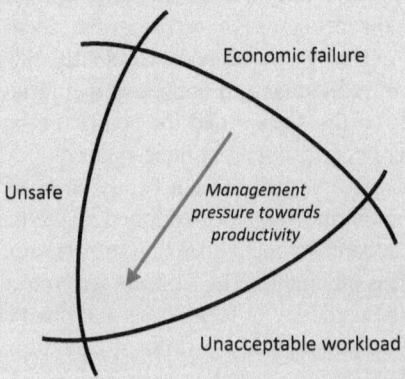

FIGURE 79   Pressure towards productivity (Adapted from Rasmussen (1997)).

At the same time, workers on the "sharp end" seek efficiency and comfort in their work by optimizing their time and effort, as shown in Figure 80:

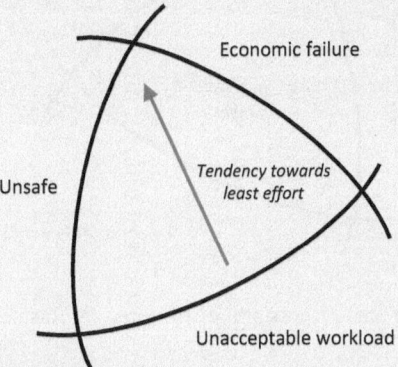

FIGURE 80   Tendency towards least effort (Adapted from Rasmussen (1997)).

The combined effect of these pressures – underpinned by the players' *loss aversion* described earlier – pushes the work towards the limit of what is acceptable in terms of safety, as shown in Figure 81, which integrates these two forces:

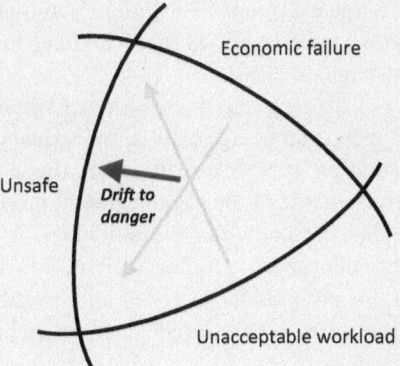

FIGURE 81   Drift to danger (Adapted from Rasmussen (1997)).

Accidents can happen when behaviour crosses this line, and this model allows us to appreciate the extent to which the causes of accidents are generally neither individual nor fortuitous. Rather, they "incubate" over long periods within an organization.

Tolerance of crossing this boundary will inevitably end in an accident, but before reaching that stage, organizations have to fight another form of tolerance: that of the *pressures underlying the migration process*. And it is only with a solid cultural anchoring of the values and beliefs that bring about prevention that this can be achieved.

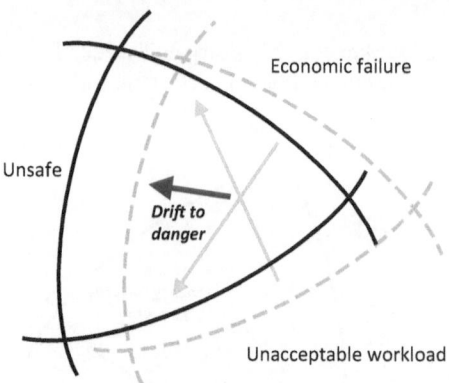

FIGURE 82   The new range of possibilities (inspired by Rasmussen (1997) and Vaughan (1995)).

Normalization of Deviance

If they do not cause any significant incident in the short term, after a while these deviations take hold and become the new norm: the range of possibilities shifts towards dangerous work (Figure 82). The concept of **normalization of deviance**, proposed by Diane Vaughan in 1995 following her investigation into the Challenger's accident described above, details the process that leads to the erosion of preventive techniques and practices and the emergence of unsafe techniques and practices within an organization that are then considered to be "normal" and are therefore accepted by the different stakeholders (Vaughan, 2016).

As with the allegory of the frog that does not react when the temperature of the water gradually rises and ends up being scalded, individuals and organizations only notice these gradual and often very slow drifts when this new state of affairs gives rise to unfortunate consequences. At the organizational level, we find the same phenomenon of habituation that we studied in the individual.

Once the drift towards dangerous work has started, it is difficult to stop it "at the right level": safety margins are gradually eroded and, as mentioned above, in complex systems this imaginary line is very rarely clear-cut and unambiguous.

Complacency

Another form of drift is **complacency**. The same mechanisms that give rise to *overconfidence* in individuals (as presented previously) can affect an organization when it manages to achieve and maintain good safety results over a long period.

Management's perception that "all is well" – perhaps based on indicators which, as we have seen, may be biased and therefore out of step with reality – in no way detracts from the *value* it places on safety, but leads to the oh-so-dangerous *belief* that the subject is completely "under control".

This may lead decision-makers to shift the focus of their attention away from prevention. And, continuing the parallel with what we saw earlier for individuals, the *inattentional blindness* that results at the organizational level eventually leads to a reduction in risk control.

Finally, even if decision-makers are as sensitive to safety as ever, the same beliefs that lead them to underestimate the risks and overestimate the efficacy of means of control in place may lead them to think that certain savings are possible, often via the *optimization by simplification* as presented in the next point.

This new *attitude* on the part of decision-makers may not involve explicit messages, but the signals are clearly *perceived* and understood by middle management, who will in turn tend to relax their focus on prevention and concentrate on the *real priorities* of their bosses. Ultimately, operators also perceive this loss of momentum despite the unchanged speech on the value that the organization gives to safety.

Over-Optimization

Before becoming a fashionable concept in construction and engineering, and then a management buzzword, **resilience** was first studied by ecologists (Holling, 1973). Defined as the capacity of an ecosystem to respond to a perturbation by resisting damage and then recovering, it is an *emergent property of complex systems* that can also be observed at the scale of individual living organisms. Evolution by natural selection has privileged this capacity to endure and regenerate, rather than deflecting aggression or avoiding disturbance at any price, as being more cost-effective.

However, resilience does entail a cost: it requires **robustness** and **in-built redundancy**. For an ecosystem or an organism to cope with disturbance without losing its identity by bifurcating to a different regime (through regime shifts also called critical transitions), its main functions need to be ensured by highly robust structures and/or by different structures that can take over if one of them fails.

Organizational resilience requires the same features. But these can be considered as dead weight by decision-makers that may be laser-focused on cost-killing. The "optimization" of industrial processes, supply chains, and payroll can be pushed too far and result in *tightly coupled systems* that become inherently brittle. The obstruction of the Suez Canal in 2021 revealed the fragility of many global supply chains, which were severely disrupted by a single cargo ship interrupting this major maritime traffic lane.

According to Charles Perrow's famous **normal accident** theory (Perrow, 1984), the combination of high hazards, complexity, and tightly coupled systems inevitably leads to catastrophe. He argued that the "design of systems should be inelegant and robust, rather than elegant and sensitive", defining robust design as "one that starts with the premise of fallibility on all parts, especially the designer", and added that "redundancies and all other safety measures should be designed in from the start and not added afterwards, since add-ons are disproportionately the source of accidents" (Perrow, 1999).

In less spectacular terms, the *lean* efforts pursued by many organizations, which result in the elimination of slack on some tasks, can have unintended consequences: for example, reducing the *muda* (Japanese for waste) related to workers' unproductive movements can result in negative effects on their health if they must remain in not-so-ergonomic postures (like sitting down, for instance).

FOCUS ON... HIGH-RELIABILITY ORGANIZATIONS

The concept of **high-reliability organizations** was developed in the 1980s to define organizations capable of operating without (critical) error for long periods of time (Roberts, 1989). Many organizations with this level of performance and operating in complex and very high-risk areas, such as nuclear power plants or air traffic control services, have been studied (Weick and Sutcliffe, 2001). Assuming the degree of uncertainty inherent in their activities and their context, and their inability to avoid all possible active failures – but unable to afford to learn by potentially catastrophic trial and error – they have in common the fact of integrating the following characteristics into their modes of operation (Reason, 2000; Weick and Sutcliffe, 2001; Laporte and Consolini, 1991):

1. A *chronic unease* with active failures, but also with latent conditions, which are actively researched and analysed on an ongoing basis to continually improve the system.
2. A *reluctance to simplify*, leading them to report, analyse, and prioritize any sign of malfunctioning and to consider any problem as potentially systemic.
3. Well-honed *routines*, based on experience acquired incrementally, for managing ordinary operations, with roles and procedures that are extremely detailed and internalized by each player (who can therefore operate safely in "System 1" within a rule-based environment), while at the same time being acutely aware of the possibility of unusual or unanticipated scenarios (requiring a switch to "System 2").
4. High *organizational adaptability* to respond to crisis situations by rapidly restructuring around key competencies, based on trust and deference to expertise.

These characteristics are necessary for the system's ability to cope with complexity, uncertainties, risks, and anomalies, but they cannot simply be decreed. They refer to both practices and attitudes and cannot be understood, let alone developed, without looking at the cultural dimension of organizations.

Safety Bureaucracy

While we can admit that systems should not be over-simplified, the opposite trend can also lead to suboptimal consequences. The proliferation of health and safety regulations in many countries and organizations has been noted to accelerate since the 1970s. Roaring litigation and public scrutiny – facilitated by mass media – have naturally and rightly induced legislators, public agencies, and corporations to raise their standards, which is in principle a good thing. Many of these rules and controls have indeed increased risk mitigation, transparency, and accountability. But pushing regulation over the top can also have deleterious effects on safety by diverting resources from prevention to mere compliance (opportunity cost).

Number of authors have criticized this **bureaucratization of safety** (Amalberti, 2001; Hale and Borys, 2013), an opinion promoted notably by Dekker, who pointed out its "unintended secondary effects" that include "a reduced marginal yield of bureaucratic safety initiatives, a difficulty in predicting unexpected events, a shift to bureaucratic accountability, quantification, and 'numbers games', the occasional creation of safety problems that result from the application of fixed rules or bureaucratic safety systems, and the real and perceived constraints on organization members' personal freedom" (Dekker, 2014).

This inflation phenomenon may result in what has been dubbed "**safety clutter**" in organizations, defined as "the accumulation of safety procedures, documents, roles, and activities that are performed in the name of safety, but do not contribute to the safety of operations" (Rae et al., 2018). In addition to the negative effects listed above, safety clutter also generates workers' cynicism in the face of "surface compliance" and frustration as it "gets in the way of getting work done". The underlying mechanism at play would be that it seems easier to add or increase safety activities – at least formally – than to officially remove or reduce them.

Risk Homeostasis

But wherever an organization may be in the cyclical swinging between simplification and sophistication – either technological or regulatory – we observe that there are psychological and social compensation mechanisms that tend to delay or even stall overall safety progress. We can use road safety again as a case study to illustrate this point.

In the 1970s, economist Sam Peltzman reviewed highway accidents and found that, although technological studies suggested that annual deaths would be 20% greater without the various safety devices legally mandated to be installed on automobiles, the gains in safety were almost totally offset by the driver response to those devices (Peltzman, 1975), a phenomenon that was confirmed by subsequent research (Adams, 1981) and which was dubbed the **Peltzman effect**.

Despite cars becoming increasingly safe and, at least in developed countries, roads being in general properly engineered and maintained, we still have too many accidents. Drivers, feeling that the level of risk is technologically mitigated, tend to be less attentive to their environment and often speed and/or engage in multitasking. However, in developing countries where cars and roads are less safe, and the rules are less present and enforced, drivers are more careful and pay more attention to the road and other drivers' behaviours – without ever being able to fully compensate for the shortcomings of the inadequate or missing technological and organizational barriers.

Following Peltzman's ideas, psychologist Gerald Wilde proposed in 1982 the **risk homeostasis theory**, which argues that every individual has an acceptable level of risk that they find tolerable for a given task or situation. He called this level of accepted risk the individual's "target level of risk" (Wilde, 1998). If a worker subjectively perceives that the hazards inherent to the job are reasonably under control their choices and behaviours will evolve to gain in self-perceived efficacy and/or comfort within the range of possibilities described above.

This range of possibilities becomes indeed wider thanks to technological or organizational improvements that push the border of dangerous work farther. But the benefits of these improvements in terms of increased safety margins will eventually be neutralized, as they are transformed by workers at all levels into gains in terms of efficacy and/or comfort.

Taking this into account, Wilde suggests that the best strategy for accident prevention should be to reduce the level of risk that workers are willing to accept, thus targeting their *attitudes* via the organizational culture as described in this book.

Uniformity

It is clear that the aim of developing safety values and beliefs, as well as the techniques, practices, and other artefacts associated with them, is to guide people's attitudes in the direction of safety, so that they adopt the expected behaviours.

However, critics could see in this approach a touch of paternalism and possibly a quest for **uniformity** on the part of management, which would be fundamentally unjustifiable from a moral point of view since it would be aimed at dispossessing workers of part of their free will without their knowledge. Uniformity would be even less defensible if, as mentioned above, we were to emphasize the diversity of rationalities as an asset for organizations.

This comment deserves a response. If, on an individual level, we make a clear distinction between influence and manipulation and only the latter is ethically untenable, it is not because the mechanisms involved are not the same: it is because the manipulator's objectives are not in the interests of the manipulated.

At an organizational level, we can adopt the same reasoning and focus on judging the morality of the cultural approach by looking at the objective pursued. Seeking to anchor beliefs, values, and artefacts that are conducive to prevention so that everyone works (whether as executive, manager, or operator) in and for safety and thus avoid accidents is undoubtedly in the interests of all workers.

Furthermore, by seeking to reinforce values and beliefs that promote safety, leaders do not deprive workers of their ability to reflect and their freedom of choice, which remain essential for carrying out activities when faced with real-life difficulties and for enabling improvement and innovation. On the contrary, it would open up the space for dialogue and get away from the monotone speech of regulated safety.

Through this dialogue, and as part of the processes of *consultation* and *participation* – to which we will return – employees can and must be involved in the process of building a safety culture.

Nonetheless, every leader in every organization has a moral responsibility to ensure that the influence they can exert on the organization and its workers through the means presented (and using the tools that will be presented later) is guided by the workers' interests. When it comes to safety leadership, it is absolutely essential to avoid using the values, beliefs, and artefacts that underpin it for other purposes (see the section on politics, below), otherwise, leadership will be undermined.

Politics

In its broadest sense, politics can encompass everything to do with the collective and how it is organized, and includes all areas of such collective: sociology, law, economics, etc. But it is politics as the practice of power that we are interested in here, understood as the tactics and strategies that individuals use to obtain what they want from and through others. All conscious human behaviour within a group where individuals have even a residue of autonomy has a political dimension.

FOCUS ON... THE PLAYER AND THE SYSTEM

Such is the title of Michel Crozier and Erhard Friedberg's seminal 1977 book. They, too, recognize the role of *attitudes* as "relatively permanent normative orientations of individuals in relation to certain privileged social objects, carved up according to centres of interest and the specific needs of research" (occupational health and safety in our case), which "refers to what is 'individual' as opposed to what is 'situational'", based "in a certain theory of social learning through the matrix of society" at that they define as "the bridge between, on the one hand, the observable behaviour of individuals and, on the other hand, the structure of values which guides this behaviour and which, in turn, remains unobservable".

The originality of their approach in relation to social psychology is that for them such attitudes, apart from being the result of this social learning, correspond to *strategic orientations* adopted by players within a group, taking into account their possibilities and the constraints weighing on them and reflecting "their choice of an action orientation in the face of the risks and opportunities of the games in which they participate in their social universes".

And in their view, these are above all **power games**, in which each player deploys adjustment strategies in relation to the others and to the system they are conforming. A power that is not synonymous with established authority emanating from the structure of the organization, but "the always contingent result of the mobilisation by players of the relevant sources of uncertainty that they control in a given game structure, for the relations and dealings with the other participants in this game".

The individual generally does not exploit these uncertainties, and therefore this power, out of meanness or malice (even if there may be a dark side of lies, manipulation, and extraction in these dealings), but because the autonomy they have to try to maintain and improve their situation lies in "their ability to negotiate their 'goodwill', their discretionary behaviour towards others" (Crozier and Friedberg, 1977).

In the relationships that are established between workers (operators, managers, executives, staff representatives, etc.) within an organization in which they carry out tasks that are never perfectly regulated, as well as between them and external players (customers, subcontractors, suppliers, authorities, etc.), there are always exchanges based on the "goodwill" of each party. Goodwill that they bargain on the basis of the relevant margins of uncertainty that each one controls in relation to the others. This is where their respective power lies.

The operator may be tempted to exploit her knowledge of the precise details of the execution of a task, including the risks, in relation to the uncertainties of the manager and/or other operators, to negotiate her wage and other forms of compensation, working rate, breaks, and other working conditions to her advantage... The manager can do as much regarding the operator, for example by hiding from her the true magnitude of the risks associated with the task or the potential improvements that could be made to her working conditions, but which would cost too much... Managers can also play this game in relation to other managers and executives. Executives in relation to managers, operators, customers, shareholders, suppliers, authorities, etc.

These power games focused on preserving autonomy, and based on the management of relevant uncertainties, can degrade communication between players and lead to a loss of mutual trust, an essential ingredient, as we shall see, for developing effective leadership in safety matters.

DID YOU KNOW? KAKISTOCRACY

It appears that it was Paul Gosnold, in a preach delivered in the 17th century, who first used the word **kakistocracy**, derived from the Greek *kakistos* (worst) and *kratos* (rule), literally meaning "government of the worst people". But recent research argues that the promotion of the "worst people" is also common practice in other institutions, such as organized crime and universities... (Alvesson and Spicer, 2016).

Those in power would be tempted to reward with appointments to key positions in their organizations not the best, but those who have shown them the most loyalty. And in particular, among these, those whom they consider to pose no danger to them because they are no more gifted than they are themselves.

Since these unfairly promoted individuals are bound to know to whom they owe their position and to suspect the reasons for it (indeed, in their case the now famous "impostor syndrome" would not be a mere illusion), they would become all the more subservient.

As they move up the hierarchy, they would follow the same logic and use their position to advance their successors through the ranks according to the same criteria, thus perpetuating the system.

In most organizations there are control mechanisms at various levels, such as boards of directors, to avoid this kind of phenomenon. But we understand by now that in intractable systems any control mechanism has its limits... An organizational culture based on positive aspirational values, such as safety, is undoubtedly one of the best bulwarks against kakistocratic drift.

Key Learning Points for the Safety Leader

Understanding the psychological processes of individuals and the sociological mechanisms of an organization does not necessarily mean that the process of integrating safety into the organization's culture will run smoothly.

The natural tendencies of organizations to drift towards danger and to normalize this deviance must be taken into account and monitored, as must the complacency that can lurk when we start to make perceptible and significant progress. All this is predictable and can be anticipated.

It is often more difficult for the safety leader to avoid the pitfalls – always contingent – of a political nature, which arise from the equally natural tendency of each individual (operator, manager, or executive) to preserve their autonomy and room for manoeuvre in relation to others to achieve their own objectives.

As we shall see, it is the development of *trust*, based on dialogue and practices that effectively open up the process to the participation of all stakeholders (from engineering to deployment, and including design and planning), that will enable us to defuse these obstacles or at least prevent them from getting in the way of prevention.

LEADERSHIP

"What you do has far greater impact than what you say".

Stephen Covey

The leader must motivate and guide the actions of others in the desired direction. As we have seen, the ultimate trigger for conscious action is **emotion**. It does not matter if this emotion is not the result of a thought process that holds water... individuals will find a way to explain their choices rationally later (a process called rationalization). Ideally, the leader should provide a vision that includes both emotion and reason, but it is the emotion that prevails!

In this light, we could think, as Plato and Plutarch did, along with many other philosophers until the beginning of the 20th century, such as Max Weber (Weber, 1914) and Thomas Carlyle with his *Great Man Theory* (Carlyle, 1888), that certain personality traits are therefore necessary to convey or generate this emotion in others, and that people with these individual attributes can naturally, through their radiance or charm, trigger this desire to follow them.

This conception of leadership became less popular in the first half of the 20th century with the emergence of other theories, before coming back into fashion with the rise of analytical psychology and the various personality analysis tools that have since been created on its principles.

FOCUS ON... FROM JUNGIAN TYPOLOGIES TO THE BIG FIVE

The theories of Carl Gustav Jung, the founder of analytical psychology, and the "**psychological types**" that he postulated (Jung, 1921), form the basis of a whole series of personality analysis methods such as the *Big Five*, MBTI, and DISC.

We will describe the *Big Five personality traits*, as it is arguably the most scientifically sound. Drawn up in 1981 by Lewis Goldberg on the basis of a series of empirical observations made in the 1950s, developed by several other research teams, and finally refined by Costa and McCrae (McCrae and Costa, 1999), it describes an individual's personality on the basis of five central traits, whose acronym is OCEAN (the name under which the model is also known):

- **Openness to experience**: inventive and curious versus consistent and cautious.
- **Conscientiousness**: efficient and organized versus extravagant and careless.
- **Extraversion**: outgoing and energetic versus solitary and reserved.
- **Agreeableness**: friendly and compassionate versus critical and judgemental.
- **Neuroticism**: sensitive and nervous versus confident and resilient.

According to studies based on this OCEAN model, individuals with high levels of openness, conscientiousness, and extraversion but low levels of neuroticism are more likely to seek a dominant position (Riggio, 2022).

Agreeableness and therefore empathy, compassion, and a spirit of cooperation do not seem to encourage people to put themselves in a leadership position, and that is a shame...

It is undeniable that certain personal characteristics have an impact on the way others perceive and judge us, but rarely do they become a real handicap to becoming a leader. In any case, their effects are marginal in relation to what the leader does and how he does it.

DID YOU KNOW? LOOKING THE PART

In North America, for the same level of education, experience, and productivity, men who are 6 feet (1.83 m) or taller earn 6% more than those who are less than 5 feet 9 inches (1.75 m) tall.

For women, everything else being equal except their weight, those over 100 kg (220 lb) earn 9% less than those under 70 kg (154 lb).

More generally, the salaries of those who correspond to the beauty standards, whether men or women, are 7.5% higher than the average, while those with an unattractive physique have to make do with salaries that are 12.5% lower than the average (Scharff, 2010).

Just goes to show that some clichés and prejudices die hard.

So, a short leader may lack that little help that nature gave to someone taller, but all in all that will not be a decisive factor. Just think of Napoleon...

In the end, it does not matter what we *believe* about personality models and theories. On the one hand, it would be impossible (or very difficult) for us to develop the traits that are by definition "natural" and that could be of interest to us. On the other hand, selecting and hiring only certain personality profiles on the basis of their supposed predisposition to play the role of safety leader is not relevant either: since we expect every worker to be able to embody it at some point, this would force us to hire only "leader" personalities, which would result in very dysfunctional teams.

Interestingly, we can also note that research has shown a significant correlation between personality traits and *values*. Openness correlates positively with self-direction and universalism values; Conscientiousness with conformity and achievement values; Extroversion with achievement and stimulation values; and Agreeableness with benevolence and tradition values. However, even if both personality traits and values influence behaviour, and that these two phenomena may be related, it appears that the influence of values is more subject to cognitive control than that of traits (Roccas et al., 2002).

So, we are going to take a closer look at the theories that make this role accessible to ordinary people. The assumption (or *belief*) is that every worker in an organization can be a safety leader, and that safety leadership can be learned and developed. You do not need to be particularly tall or have a deep voice or square chin, a strong personality, or charisma, just as you do not need a title, hierarchical position, or specific function within the organization.

Some of the tools we will be referring to have already been introduced in previous sections, others have just been mentioned and will now be explained in detail, and some new ones will also be proposed.

LEADERSHIP STYLES AND SITUATIONS

Ensuring the safety of the workers over whom they have authority is an inescapable duty of anyone with management responsibilities. And this implies that managers need to assume a leadership role in this area. But these are not the only people who can exert a positive influence as safety leaders.

How do we approach this task so that we can carry it out effectively, whatever our role within the organization? Let us look first at the question of leadership style.

Leadership Styles

Whether it is a question of *personality* (about which we have nothing more to say because there is little we can do to change it) or *attitude* (about which we have already said enough), different leaders demonstrate different **leadership styles**. These styles have been studied in various ways by looking into different parameters, but most often they are classified according to their position in relation to continuums presented in the form of axes.

Several warnings are necessary. Firstly, some leadership styles may appear more "appealing" than others to different individuals (depending on their values and beliefs), but, as we shall see later, it can be argued that none of them is inherently more effective than others, their relevance depending on the situation.

Secondly, when it comes to the leadership styles that these individuals actually deploy, all you can get by asking them are their leadership *preferences*, based on their spoused values and beliefs, which are often more linked to the operating modes that they reckon are more relevant than to those that they implement spontaneously and/ or most effectively.

Finally, like the criteria presented above for the study of psychological types, all these dichotomies contain a significant element of Manichaeism and imagination. A bit like with zodiac signs, if one does not have a very good factual basis through direct observation of the individual's behaviour in real situations and a lot of perspective, one can always build a story to classify someone (or oneself) in a given category. If not in relation to their date of birth, this classification can be influenced by the subjective image that one has constructed of them according to one's perception and local rationality, which is always partial and influenced by affects and one's mental models (or one's own aspirations).

The point of looking at this literature is not to find the ideal leadership style, or even the best taxonomy or "catalogue" of these styles to classify leaders according to their traits so that they can be matched with the right mission. The objective is rather to obtain indications of how leaders can adopt and deploy the practices that would appear to be the most appropriate in various situations, contexts, and cultures, practices to which we will return a little later (Likert, 1967).

Here is a brief summary of some of the axes used to classify **leadership styles**, without any claim to exhaustiveness or guarantee as to their scientific rigour. All we will do, very briefly, is to compare these leadership styles with the "cultural styles" (if we dare call them that to emphasize the parallel in the approach) derived from the classification based on the axes defined by Hofstede and presented above:

1. **Authoritarian (or autocratic) versus participative (or democratic)**: Whether their authority is based on hierarchical position, seniority, skills, or the influence they exert over others, *authoritarian leaders* try to have a monopoly on decision-making and control the group, placing themselves above it. At the other extreme, the *participative leaders* encourage autonomy, initiative-taking, and decentralized collaboration within the group, of which they consider themselves to be members like any other, with the role of facilitator (or "servant" in the most radical version, *servant leadership*).

   Between these extremes we find the *paternalistic leaders*, who ensure benevolent control by giving priority to the collective interest as they see it, and the *consultative leaders*, who reserve the decision-making role for themselves but seek everyone's opinion.

   In relation to Hofstede's classification of cultures, authoritarian leadership would be appropriate for those characterized by a high power distance (PDI), whereas in those with a lower PDI, participative leadership would be more fitting.

2. **Directive versus delegative**: At one extreme of this axis (which is similar in some respects to the previous one), the *directive leaders* impose objectives and put in place procedures and controls to provide a clear framework for the decision-making process. In contrast, the *delegative leaders* practice laissez-faire, giving the team freedom not only in the choices they make but also in the decision-making process they deploy.

   Directivity would be required in cultures with high PDI and uncertainty avoidance (UAI), while delegation would be apt in cultures with low PDI and UAI.

3. **Relational (or affiliative) versus organizational (managerial)**: Above all, *affiliative leaders* seek cohesion and harmony within the group and ensure the quality of relations between its members by establishing a climate of

trust. The *managerial leaders* correspond to the archetypal bureaucrat, who define rules and assign responsibilities, then set about supervising compliance by the individuals in their charge, understanding harmony as the absence of problems that should logically ensue.

The first would be better suited to cultures with a lower UAI, collectivist, cooperative, and indulgent, and the second to cultures with a higher UAI, individualist, competitive, and characterized by restraint.

4. **Coaching versus challenging**: *Coaching leaders* aim to identify the strengths and weaknesses of others so that they can best deploy the former and support them in their development to make up for the latter, which in the long term not only improves performance but also the overall competence of the team. *Challenging leaders* set ambitious objectives, often with a short-term focus, then put individuals and the team under pressure to mobilize energies to the maximum through successive sprints.

The former would be appropriate in cultures with a low PDI and a high UAI, collectivist, cooperative, and long-term oriented, while the latter would be suitable in cultures that are individualist, competitive, and short-term oriented.

5. **Front runner versus strategist**: The *front runner leaders* are directly involved in day-to-day operations and "sweat their shirts". They are the captain who leads the assault on the front line or the supervisor who arrives first on-site every morning. The *strategic leader* is more concerned with analysis, anticipation, and steering.

The former would be better suited to cultures with a low PDI and a high UAI, while the latter would conform better to cultures with a high PDI and a lower UAI.

6. **Visionary (or inspirational) versus pragmatic (or reassuring)**: Like strategic leaders, *visionary leaders* are in their element when it comes to strategic thinking, but they are more interested in foresight, innovation, and inspiring teams and less in following up actions. *Pragmatic leaders*, like front runners, are action-oriented and problem solving, but unlike the front runners, their main objective is not to do things and show people how to do things but to ensure that they have the right information to make the right choices.

The first would be better suited to cultures with a low UAI and a long-term focus, while the second would rather belong to cultures with a high UAI and a short-term focus.

7. **Transformational versus transactional**: A blend of coach, front runner, and visionary, the *transformational leader* not only develops the vision but also ensures that others share in it by making tangible efforts to gain their trust and support them in adopting it. The *transactional leader* is somewhere between the challenger, the strategist, and the pragmatist: analytical and acting as a pilot as well as political and very down-to-earth, seeing motivation as a bargaining exercise.

The first would be more successful in cultures with a low UAI, collectivist, cooperative, long-term oriented, and indulgent. The second would rather fit in cultures with a higher UAI, individualistic, competitive, short-term oriented, and which favour restraint.

Situational Leadership

As we have just seen, it can be argued that a leadership style may be more or less likely to succeed depending on the culture of the organization (and the underlying macro-cultures). But within the same cultural framework, the success of a given leadership style will also depend on the specific situation that the leader has to deal with. So ideally, the leader should not be the prisoner of one style but should be able to analyse the situation, decide on the most appropriate leadership style, and apply it on the spot. This is the philosophy of **situational leadership**.

It diverges from the trait theory as it assumes that individuals, and therefore leaders, are capable of adapting their behaviour, even if it admits that, because of idiosyncrasy, conviction, or habit, everyone tends to follow a style that is more natural to them and that it takes a deliberate effort for them to depart from it.

Various authors have proposed different factors to characterize the situations leaders face and determine the type of leadership that would be most effective in them.

FOCUS ON... SITUATIONAL LEADERSHIP MODELS

Situational Leadership is not only a registered trademark but also, and above all, a theory whose origins date back to the 1940s.

FIRST ATTEMPTS

It was during that decade that the theory of traits began to include indications of the circumstances in which each leadership style produced the best results. For example, based on Kurt Lewin's experiments on group dynamics and the three types of leadership he identified, the following correspondence was established (Lewin, 1939):

- Directive leadership: adapted to *crisis situations*, to control and reassure.
- Participative leadership: adapted to *situations requiring consensus*, to encourage commitment.
- Laissez-faire leadership: adapted to *stable periods*, to encourage innovation and individual development.

CONTINGENCY MODEL

In the 1960s, Fred Fiedler proposed three main variables that determine the effectiveness of a leadership style in each situation (Fiedler, 1964):

- *Trust within the team*, based on the quality of relationships, which would be the most important parameter (to which we will return later)
- *The structure of tasks*, ranging from low structure (ambiguous procedures) to high structure (clear procedures)
- *The legitimate power* enjoyed by the leader, enabling her to impose, reward, and coerce.

THE MATRIX APPROACH

In the 1950s, a team from Ohio State University identified the need to combine some of the parameters that had previously been studied separately. They constructed a matrix to study leaders' behaviour, with *structuring* (the desire to structure roles to achieve an objective) on one axis and *consideration* (the desire to establish mutual trust) on the other. The following decade, another team from the University of Michigan proposed a similar matrix, but with one axis focusing on *relationships* within the team and the other on *production*.

Robert Blake and Jane Mouton built on this work and, presenting an interest in people on one axis and interest in production on the other, they constructed a grid with 81 styles, of which they described the five main ones in detail (Blake and Mouton, 1964).

OBJECTIVE – TRAJECTORY AND MATURITY MODELS

Taking this matrix idea a step further, Paul Hersey and Kenneth Blanchard constructed their own matrix based on the four leadership behaviours identified by Robert House (delegative, directive, participative, and supportive) in which they integrated four degrees of employee maturity (resulting in turn from the combination of their knowledge and motivation) and established the following correspondence (Hersey and Blanchard, 1969):

- M1 – Low maturity (employee combining low knowledge and low motivation) → requires directive leadership
- M2 – Medium-low maturity (employee combining low knowledge with high motivation) → calls on supportive leadership
- M3 – Average maturity (employee combining high knowledge with low motivation) → calls for participative leadership
- M4 – High maturity (employees combining a high level of knowledge and motivation) → calls for delegative leadership.

On the basis of what we saw in the previous chapter about the knowledge and motivation of individuals, the objective – trajectory and maturity models can provide safety leaders with an interesting key to adapting their behaviour to the situation they need to address.

The safety leader must analyse, on the one hand, the worker's degree of *knowledge* of the risks involved and the techniques for managing them (whether formal instructions or informal practices, which we shall call **competence** as previously defined) and, on the other hand, their motivation to adopt them (which we can link to their **attitude**):

- When facing workers who take risks because they are unaware of them and/ or have not mastered the techniques for carrying out the task safely, and who also show little motivation to adapt their behaviour when they are told what they should do (**M1**), the safety leader must give precise instructions, ensure that they are taken into account and take a firm stance in the event of deviation, not hesitating to apply the discipline schemes as mentioned above.

- Faced with workers who take risks because they are unfamiliar with them and/or have not mastered the techniques for carrying out the task safely, but who are genuinely open to feedback and inclined to adopt the right behaviour (**M2**), the safety leader must provide the necessary guidance and instructions and be available to clarify any misunderstandings. Resulting deviations would probably be due to a lack of skill, competence, or precision in the instructions given, and not to inappropriate attitude, therefore calling for understanding and pedagogy.
- Faced with workers who take risks even though they have perfect mastery of the task and the techniques they should be applying, but who are ostensibly reluctant to do so (**M3**), the safety leader must be prepared to listen to their arguments and be open to their ideas, inviting them to help in the process of optimizing the applicable rules to reduce the constraints on them, if this can be done without compromising safety. Workers in this quadrant are often the most difficult to manage, as they are experienced and skilled. They will show little interest in prevention arguments, when they are not outright hostile to them, and always have good reasons to do what they do in the way they do it. This advice to leaders to remain open and to listen to these reasons, even though they may seem obviously emanating from local and bounded rationality and from values that are not consistent with the organization's safety culture, may seem counterintuitive. In the Toolbox (Tool No. 11) we present a technique to facilitate these exchanges: the behavioural-specific feedback.
- When faced with suggestions from workers to reduce or waive certain safety constraints, if they are fully conversant with the risks associated with the task and the applicable techniques and have systematically demonstrated their adherence to them (**M4**), the safety leader should encourage them to think about how to formalize the adjustments they propose and only intervene to assist them if necessary in this process, as well as at the end to review them and validate that they guarantee a satisfactory level of safety (i.e. in line with the values and beliefs of the organization). This leadership behaviour not only recognizes experience and rewards positive safety attitudes but is also an excellent way to "declutter" safety in some instances.

Again, this approach is situational, not personal (which would bring us back to trait theory) (Figure 83). The same worker may be at the "M2" stage for a given task and simultaneously at the "M4" stage for another task, in relation to which she would have accumulated considerable expertise. Above all, and this is why we talk about maturity, for the same task a worker would logically move over time from the "M1" stage to the "M4" stage.

With regard to this point about the progression of maturity, as much as the evolution of skills and competence over the course of a worker's career seems logical, it might also seem counterintuitive that her motivation should be low at the start, then

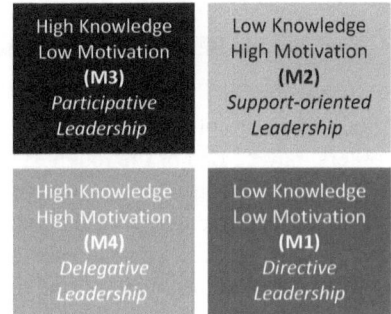

FIGURE 83 Situational leadership. (Adapted from Blanchard)

increase, then fall, only to rise again. Various authors give different explanations for these comings and goings, but they are of no importance to us: what matters is the need to identify the workers' current state of motivation (which could be likened to the worker's safety attitude) and combine this information with their degree of competence and thus be able to determine the best way of approaching them.

Key Learning Points for the Safety Leader

There's no denying that everyone has their own personality and character, and that relationships between two individuals are more fluid when they are "compatible".

Safety leaders are also entitled to their temperament, but their prevention-related behaviour will transcend these traits, being driven by an attitude based on values and beliefs that promote safety.

It is this attitude, systematically aligning their own behaviours in the direction of safety, which makes them legitimate when they address others to try to guide theirs in this area, whether with a remark, a suggestion, a compliment, or a reprimand.

Safety leaders can still make mistakes in their interactions with other workers if the former do not adapt their approach to the maturity of the latter, typically:

- If they fail to take account of their *competence* (skills, knowledge of the task, and the risks) and/or
- If they fail to correctly assess their motivation to adopt the expected behaviour (usually closely linked to their attitude towards safety).

It is therefore vital to analyse these parameters to adopt the most appropriate leadership style in each instance.

TOOLBOX – Tool No. 11: Behavioural-Specific Feedback

"This is the way I've always done it", or even worse, "*We*'ve always done it like this". Addressing individual or collective unsafe behaviours executed by highly experienced workers may be one of the toughest challenges for safety leaders. According to Hofstede's model of cultural appropriation previously described, these workers are less prone to evolve in their core values, beliefs, and working techniques. In the long run, we can expect to slowly orient their attitudes as we grow safety-related values, beliefs, and practices at the centre of the organizational culture, with the tools we will see later.

In the short- and medium-term, we can now complete the PIC-NIC technique that we presented in Tool No. 4 with this other method for delivering the most effective feedback. It has been taught by David Bradford (Bradford and Robin, 2021) in his Interpersonal Dynamics course at Stanford for several decades and, although he did not explicitly make the link, it appears to be tailored for these workers sitting in the Situational Leadership's M3 quadrant: those with high knowledge of risks and what would be the safest way to do the task but low motivation to go through the motions of safe working techniques. The recipe can lead to dangerous overconfidence.

Years of first-hand experience without severe accident have consolidated their attitude towards risk-taking, and all the biases we have presented (loss aversion, availability heuristics, status quo, expert bias, confirmation bias, conformism…) make them impervious to the traditional managerial approaches, ranging from patiently and paternalistically lecturing them to yelling and threatening.

As we stated above, the leader remaining open to dialogue has been demonstrated to be the best strategy to defuse the defensive mechanisms that will naturally be opposed by these experienced workers when their know-how is challenged. Let us remember that, in their psyche, questioning their working methods and the underlying assumptions is tantamount to questioning their professional identity and worth.

How to start and conduct this dialogue to make the most of it? The essence of **behavioural-specific feedback** as proposed by Bradford is simply "sticking with your reality" when considering the different areas of understanding (or "realities") that exist whenever a safety leader and a worker interact, regardless of their respective status. There are no less than three realities at play:

1. The *worker's intent*, based on her needs, motives, emotions, and self-perceived situation, that we can relate to her *attitude*.
2. The *worker's behaviour*, that is visible to both the worker and the safety leader.
3. The *impact of the worker's behaviour on the safety leader*, the interpretation that the safety leader makes of the worker's behaviour as the leader perceives it, and what that makes the leader figure and feel based on her own *attitude*.

The key to a productive dialogue would then be for the safety leader to avoid making assumptions about the first reality (thus forgoing what we presented about the theory of mind), which belongs to the worker. The safety leader shall only point out and comment on the second reality, the observable behaviour, and the third one: what the behaviour as she perceived it makes her think and how it makes her feel. This kind of feedback is indisputable.

Bradford illustrates this dividing principle with the net that separates the two sides of an imaginary tennis court: the first reality is on the worker's side of the court, and the safety leader should not jump over the net by making assumptions about it. The moment she does so, she commits a fault, and the dialogue is over.

The safety leader must, on the contrary, interrogate the worker using open-ended questions to unveil the perception, needs, motives, and emotions that brought the worker to act as she did. If and when this is done in a non-judgemental way, by displaying true curiosity, we can expect the worker to honestly engage in the dialogue that is proposed to her, sending back the ball instead of closing up in defensiveness.

Maybe the safety leader did not appreciate the situation correctly? The many perceptual limitations we have presented apply to safety leaders as well, and acknowledging our mistake would be all the less embarrassing if we offered our feedback as a humble inquiry. Maybe the worker's rationality holds water? The safety leader, and the whole organization, can then learn something and eventually adapt the rules and procedures.

If not, it will be up to the safety leader to explain why the reasons opposed by the worker are not valid and why the rule (or the organization's "code of ethics") must be followed. Most of the time this will not satisfy the "M3" worker, but at least she has been given the opportunity to explain her behaviour. By doing so, she may get to re-examine her rationality and eventually reconsider her beliefs about the magnitude of the risk she was facing and the efficacy of the risk mitigation she was deploying. In any case, and above all, she has been shown respect, and that makes all the difference.

Like the other tools in this book, behavioural-specific feedback is not a magic wand that will instantly light up the mythical light bulb in the worker's head, but if used consistently it offers the safety leader a decent chance of making progress in orienting the worker's attitudes towards safety.

EFFECTIVE LEADERSHIP

Once this important point about leadership styles has been made, we can move on to the subject of **practices** and other **leadership artefacts** and how to apply them to anchor and develop safety in the culture of organizations.

We have already talked about the two main tools that leaders have at their disposal to shape the culture of an organization: *setting an example* and *adequate feedback*. In this section, we add to those three series of tools, some of which further develop these two themes.

The first two series of tools were proposed by Edgar Schein, who distinguished between *primary mechanisms* for integrating values and beliefs and *secondary mechanisms* for reinforcing and stabilising them (Schein, 2017). The former are *practices*, the latter are *other types of artefacts*. Taken together, they form an initial toolbox for developing beliefs and values within an organization, so that its culture can evolve in the desired direction. They are presented through the safety lens and in decreasing order of importance, but without any temporal sequential logic because they can all operate simultaneously.

The third is based on the recommendations made by management professor Amy Edmonson, who, after studying numerous organizations, proposed the main keys for developing *psychological safety* (Edmonson, 2019).

Other authors and institutions have presented safety leadership approaches on similar lines, recombining some of these elements, but in the end we always find the same basic ingredients (ICSI, 2022).

Primary Embedding Mechanisms (Practices)

As mentioned above, the precise way in which the safety leader applies all the practices below will depend on the culture of the organization and the underlying macro-cultures, as well as on each leader's preferred style. But apart from these adaptations, on the whole they are universally valid and constitute the ABCs of leadership, which we can (and must) apply to prevention.

1. Systematic, daily focus on prevention

The first and most important thing a leader can do is to focus on safety. From simple comments to rewards or punishments (see point 5 below), each behaviour of a leader, not only what they notice and what they do but also what they do not notice and what they do not do, demonstrates their own beliefs about prevention and the importance they give to this subject, and therefore their values.

This point was developed in the chapter on *attention*: it is spontaneously drawn to what matters most to us, and others are quick to notice what our attention is systematically focused on. Even if one-off intensity can have a certain impact (the "outburst"), it is above all the consistency and continuity of these behaviours that have an effect. This is the principle of setting an example and demonstrating personal commitment.

We can all forget a birthday, even if it is very important to us, when we are swamped with emergencies... But if this unfortunate oversight is repeated several times, the person concerned may legitimately question the importance that this event really has in our eyes, and by extension, the importance that we give to our relationship with them.

The same principle applies to safety in an organization. A leader may pass by a dangerous situation without noticing it once, her head being elsewhere that day. But if this happens systematically, her teams will quickly understand what her values are and what is really important to her.

The same coherence that is necessary in a leader is also essential throughout the organization. All leaders must fully consider the value of safety when it comes to making trade-offs (Figure 84).

It's not always easy to follow leadership...

FIGURE 84 Leadership (drawing by author).

2. Reactions in the event of an accident or crisis

The behaviour of leaders during serious events and crises is particularly striking for the team members. It is in these moments that we reveal the true hierarchy of our values: if, following a dramatic accident, the leader's main concern is how to manage the impact in the media, it will be very difficult for her to go back on that impression.

Especially since, as we have seen, during these emotional moments the learning for those watching is particularly intense. Reactions and behaviour during these situations will therefore be all the more significant.

In every organization there will be at some point a serious safety event, whether involving an employee, a sub-contractor, or a customer, in the course of business or during commuting. Leaders must be prepared to deal with these situations, and their "crisis management plans", whether they are perfectly formalized or not, must include the management of the emotional component.

The sincere demonstration of compassion and empathy, as well as the willingness to learn and improve without shirking responsibility, will reinforce the value of safety within the organization.

3. Allocation of resources

The priority (or lack of priority) that is systematically given to prevention when deciding on the resources to be allocated to it, typically when drawing up budgets, reflects the beliefs and values of leaders in these areas, which go so far as to limit their ability to identify alternatives.

But it is not just budgetary processes that count. All workers, from decision-makers to operators, have discretionary margins to direct resources at their own level (if only the time and energy they devote to safety-related issues).

The safety leader mobilizes the resources under her control or over which she can have an influence to ensure that prevention-related issues are managed diligently.

4. Deliberately adopting the role of model, teacher, and coach

Leaders are generally aware of the importance of getting the right messages across at assemblies, town hall meetings, training sessions, and other rituals such as external "public relations exercises" or internal "indoctrination sessions".

However, it is above all the messages that workers receive on a daily basis (verbal and non-verbal, explicit and implicit) that will determine how they perceive the true values and beliefs of their leaders, and therefore whether they will adopt them or not.

Among their duties, safety leaders must give priority to coaching others to pass on good safety techniques and practices. The behavioural-specific feedback technique presented in the Toolbox (Tool No. 11) can help leaders to adopt the right tone. But beyond mere "corrective feedback", safety leaders must remember that one of their main objectives should be to encourage other safety leaders to emerge, as we will develop later.

5. Rewards and punishments

Leaders must systematically provide feedback tailored to the behaviour observed (again, refer to the tools already presented), whether or not damage has been caused. Rewards and punishments must, we repeat once again, be understood in the behaviourist sense of these terms: they are not limited to bonuses and penalties but encompass a whole range of gestures which are generally subtle, even subliminal, but that have a very important psychological impact in terms of recognition, commitment, and motivation.

As already mentioned, safety leaders (especially if they occupy a position conferring them hierarchical authority) must be fully aware of the impact of their feedback on the teams and, consequently, make every effort to avoid feedback that could be perceived as partial, biased, or arbitrary.

Obviously, this principle should not be limited to the behaviour of operational staff, but must also apply to the choices and decisions made by managers and executives, and should be consistently applied to all formal appraisal and incentives, as well as promotion and dismissal procedures (see point 6 below).

6. Selection, promotion, and exclusion

As already commented, organizations select new members and promote or exclude old ones by taking into account, in a more or less deliberate way, their cultural compatibility. The organization's understanding of the beliefs and values it seeks to develop is essential to optimizing these processes.

The symbolic significance of these mechanisms and their ability to promote safety-related values and beliefs can be destroyed if they are misused by political abuses such as those described in the section about drifts.

Secondary Reinforcement and Stabilizing Mechanisms (Other Artefacts)

Most of the tools presented in this second list are not practices, but rather other artefacts as we have defined them, i.e. organizational "objects" (whether material or immaterial) whose purely practical function is accompanied by a symbolic function, and which are therefore imbued with cultural significance.

1. The structure of the organization

The formal structuring of the safety function, its reporting lines, and its interfaces with the other functions within the organization are all ways of expressing the beliefs and emphasizing the value that the organization assigns to prevention.

For example, the fact that the safety function reports directly to senior management does not in itself ensure that these aspects are taken into account more effectively, but it does show a degree of prioritization. If it does not report to the CEO or GM, its attachment, for example, to an HR department or an Operations department may also create different perceptions and expectations among the people it deals with, who may consider it more legitimate a priori in matters relating to skills, training, and communication in the first case, or in technical and operational matters in the second.

Many organizations have matrix structures with hierarchical links on one side (the solid line) and functional links on the other (the dotted line). If the staff in charge of these preventive activities report hierarchically to the head of the safety function (and in a dotted line to the person in charge of the activity), we can imagine greater independence from operations, but we can also fear an approach that is more oriented towards prescription and control. Conversely, a hierarchical reporting line to the operational manager (and in a dotted line to the safety manager) often entails closer collaboration with operations, but at the cost of a subordination that can be problematic in the event of divergent points of view or conflicts of values.

Whether formally considered as such or not, the safety function is always a support function. It very rarely has total autonomy or real authority over those in charge of operations. Thus, beyond the practical aspects, the symbolic authority conferred by its reporting line can be significant.

2. Systems and procedures

Management systems and the procedures and other documents that make them up formalize the systematic, day-to-day focus on prevention (see point 1 in the list of practices above). They detail the techniques and practices which the organization has identified as having to be systematically implemented to ensure prevention. Standards such as ISO 45001 provide a framework, listing the essential elements and the guides for managing them, to ensure a systemic approach as we have described.

The way in which these elements are documented and integrated into the overall landscape of the organization's management systems and tools is far from neutral: if they are done in a "professional" way, they will be viewed with more consideration than if they reflect a certain amateurism (for example, work done by a trainee on the eve of the certification audit). Coordinated work with the quality, engineering, and internal communications functions, to name but a few, will be essential at this level.

3. Rites and rituals

Toolbox talks, dynamic risk assessments, and induction training for new recruits, as well as incident analyses, management reviews, and other audits, are examples of ritual activities whose practical purpose is coupled with a symbolic function.

The inclusion of certain actions, statements, and sequences dedicated to prevention in other rites or rituals can also have an important symbolic impact, such as the systematic presentation by the organizer of instructions for evacuating or making the room/building safe before starting any meeting or training course with participants from outside the site.

The meaning they convey makes them double-edged tools. When they are well conceived, well deployed, well executed and therefore perceived as useful, they reinforce the value and beliefs associated with safety. However, when one or more of these conditions are not met, workers perceive them as mere impositions, or worse, as constraints that waste their time and distract the organization's attention and resources from the real problems.

In this second case, they will still be symbols, but of the insincerity of the organization's commitment to prevention. Instead of reinforcing the value it places on safety and the beliefs that go with it, we obtain the opposite result.

The design and facilitation of these rites and rituals must, therefore, be entrusted solely to genuine safety leaders, underlining the importance of the process of generating these leaders, the subject of this book.

4. The design and layout of the premises

Like certain rites and rituals, the signposting of hazards and obligations, visual standards, information systems, and the very layout of workplaces and equipment can emphasize prevention or, on the contrary, reflect the organization's lack of interest in the subject.

Safety leaders need to get involved in these aspects, going beyond the traditional role of the safety practitioner, which can be imagined as being limited to identifying potential hazards and proposing and/or validating technical options for managing the risks. They must work with the design teams at a very early stage to ensure that prevention is taken into account in every choice, by integrating all the human factors and cultural dimensions that we are discussing in these pages.

5. Stories about important events and people

From the legend of the visionary founder to tales of great commercial or industrial triumphs, every organization puts forward, in a more or less planned and coordinated way, stories that convey its values and beliefs. The power of storytelling to transmit, solidify, and perpetuate these cultural pillars, and to inspire and motivate individuals, is recognized and used (even over-exploited) in many fields, particularly in advertising.

In the area of prevention, these are often stories of serious or even catastrophic events that are passed on to newcomers or mentioned on occasion, for example to remind of the importance of adhering to a particular practice. As we said when we looked at the subject of memory, stories with an emotional component are better remembered and therefore more likely to be mentally available when it comes to assessing a risk. The safety leader must therefore know how to use this tool.

6. Charters, creeds, and other formal statements

When they are not just a formality, and when the managers of an organization, as enlightened leaders, carry out a genuine exercise of introspection to bring out the true values and beliefs on which they wish to anchor the organization's culture, these messages are significant in articulating its identity.

If these conditions are not met, despite the symbolism of the beautifully framed commitment with the signature at the bottom of the page and displayed in the entrance hall and/or Internet website, they are nothing more than propaganda tools whose only use is external communication.

The difference between "external communication" and "internal communication" is that individuals within an organization can compare the messages they receive with the reality of the practices they observe and the other artefacts actually implemented. Let us reiterate once again that "communication campaigns" can only serve as a support for the development of a culture, provided that real organizational and managerial efforts are made to align the management practices and other artefacts in the desired direction.

Since all contributions are welcome, if one wishes to use this lever too, we must remember what we said when we talked about perception: priority to graphic and multimedia information. Reading involves higher cortical processes, while vast areas of the brain are dedicated to processing visual signals.

Consultation and Participation

To the mechanisms (practices and other artefacts) listed above, we propose adding those needed to involve workers at all levels by ensuring that they are consulted and promoting their active participation in safety matters.

As discussed above, these practices are associated with the participative leadership style which is in principle more typical of English-speaking countries where distance from power (PDI) and UAI are relatively low, and even more so in Northern Europe, where these traits are combined with a strong spirit of cooperation.

The question therefore arises as to whether it is possible to change the culture of prevention by means of these consultation and participation mechanisms in countries where PDI and UAI are high and cooperation is low. Are consultation and participation universal levers of leadership? Or are there macro-cultural contexts and/or organizations in which they bring no benefits or are even counterproductive?

Experience tends to show that, irrespective of where the cursor is placed on the organization's macro-culture in the PDI and UAI axes, the development of these consultation and participation practices will be positive. In some cases, they will start from further back and will probably never achieve to integrate workers' participation as a *core value* of the organization, but this is neither an objective in itself nor a sine qua non for developing a safety culture.

It goes without saying (but perhaps it is better to say it) that these consultation and participation initiatives must emanate from a sincere desire to listen, exchange views, and engage in dialogue, and not be limited to asking for opinions that would be skimmed over in a paternalistic or even hasty manner. Initiating these processes means making a commitment to trying to understand the rationale behind every comment, every question, and every idea expressed. Not just to provide an answer, but with a willingness to use it as a starting point for developing their full potential together.

As we said earlier, as long as workers have even a minimum of autonomy (which is always), their behaviour and their relationships within an organization have a political dimension. If the organization does not establish a framework within which this dimension can be "broken down" as transparently and effectively as possible, these aspects will poison relations between the various parties.

Before even referring to the ethical (and therefore cultural in the sense mentioned above) argument which could justify the preference for developing relationships based on trust rather than fear because of the value attributed to it, a number of studies (certainly of English-speaking origin) have shown that the former is a superior factor in motivation and performance than the latter (Zac, 2017).

Perhaps the most effective safety leadership over the long term would be that which moves away from the messianic leader–follower model and succeeds in developing trusting relationships based on mutual respect. And that means listening actively and genuinely to create a culture where everyone can express their ideas, raise their fears, and give their opinion (Schein and Schein, 2018).

DID YOU KNOW? PARTICIPATION AND BEHAVIOUR

The famous Australian psychologist Elton Mayo, one of the pioneers of the sociology of work and the founder of the *human relations* movement in management, emphasized the importance of worker autonomy and participation in motivation (it was he who directed the experiments carried out at the Hawthorne Works factory mentioned earlier).

Subsequently, Kurt Lewin developed in greater detail the productive advantages of these components of autonomy and participation, over and above simple recognition. During the Second World War, at the request of the American army, he carried out experiments to find a way of persuading the population to consume offal, so that the meat could be sent to the troops (Lewin, 1947).

While in the control groups a nutritionist gave a simple course on diet, scarcity, and patriotism, in the experimental groups, after the same course had been given, the housewives got together to discuss what they should do.

When the results were analysed, it was found that those who had taken part in the experimental groups, and therefore actively participated in the joint reflection, changed their behaviour much more than those in the control groups.

This is the major evolution in the 2018 version of the ISO 45001 standard compared with its ancestor, OHSAS 18001. While this standard incorporates the notion of culture for the first time and highlights the role of leadership as the driving force behind the continuous improvement loop, it is the new sections dedicated to consultation and participation that provide genuinely new tools (within the standards framework) for fuelling it with trust.

Regulations in many countries have long imposed clear obligations on employers in terms of information, consultation, and participation, as guidance documents at the multinational level (EU-OSHA, 2012) have long highlighted the need for dialogue between the various stakeholders (employers, employees, and staff representatives) within organizations. Unfortunately, in most macro-cultural contexts, it is rare to find truly constructive partnerships in these formatted exchanges between employers, unions, and safety committees.

FOCUS ON... HIGH CONTEXT VERSUS LOW CONTEXT COMMUNICATION

Before initiating or reviewing consultation and participation processes, and more generally any form of exchange and dialogue within an organization, it is important to elucidate the way in which communication is structured in the macro-culture in which it operates.

Problems of understanding and interpretation often arise between individuals who do not share the same symbolic references (artefacts), beliefs, and values. But communication itself is structured differently depending on the cultural context.

In some cultures, messages are supposed to be very explicit and clearly formulated, while in others the message is more subtle or even cryptic (Hall, 1959). A continuum has been described in this message contextualization dimension (Meyer, 2014).

The United States, Australia, and Canada, and to a lesser extent the Germanic and Nordic countries, occupy the so-called **low context** side: in these countries, the receiver of a message expects the sender to include in their message all the information necessary for its meaning to be fully understood.

Asia, the Middle East, Africa and, to a lesser degree, countries such as Russia and France are on the opposite side of the spectrum, known as **high context**: in these countries, the message is conveyed as much by the information delivered by the sender as by the implicit references to cultural elements common to both parties.

A person used to "low context" communication will find it very difficult to understand a message from a "high context" speaker. Conversely, an individual from a "high context" culture may feel infantilized by "low context" communication. Difficulties in understanding can also arise between two people who are used to communicating in a "high context" but who come from different cultural backgrounds.

Key Learning Points for the Safety Leader

The 14 artefacts presented above complete our Toolbox. They have proven their effectiveness in providing lasting momentum to the cultures of many organizations, irrespective of the macro-culture in which they are embedded and the leadership styles of those who use them.

Their ongoing application to safety makes it possible to place the values and beliefs associated with it at the heart of an organization's culture and, in so doing, to guide the behaviour of its employees.

The first 12 are listed in descending order of impact. There is no chronological order to their deployment, which may be simultaneous. The last two can be added when the minimum trust conditions are met, as explained in the next section.

TOOLBOX – Tool No. 12: The Leader's "Other Tools" for Developing a Culture

Below, in the form of a checklist, we summarize the leader's "other tools" for making safety a core value in an organization over the long term, in addition to the two basic tools presented earlier (exemplarity and feedback).

For each of these tools, whether practices or other artefacts, safety leaders must ask themselves the following questions:

- **What am I already doing/what have I already put in place to achieve this? Am I doing it systematically/is it actually in place and effective?** What obstacles do I encounter and how can I remove them? How can I ensure that what works is sustainable? Can I develop it further and expand it?
- **What am I doing that should be stopped?** What have I put in place that should be scrapped (because it's ineffective or downright counterproductive)?
- **What else could I do/put in place?** What "best practices" could I learn from (in other organizations, in my sector of activity or not)? What resources would I need (time, sponsorship, material and human resources for design, development, implementation, steering, etc.)?

This questioning needs to be first of all carried out at a personal level, but then safety leaders also need to ask it:

- At the level of their scope of responsibility and/or influence within the organization, then at the level of the organization as a whole insofar as they can hope to have an impact, even if this is apparently minimal and/or long-term.
- With regard to other leaders in the above-mentioned areas, whether or not they have hierarchical authority.

Checklist of "other safety leadership tools":
Primary embedding mechanisms (practices)

1. Systematic daily focus on safety
2. Appropriate responses to accidents and crises
3. Strategic allocation of resources to safety

4. Deliberate safety role modelling and coaching
5. Judicious allocation of rewards and punishments
6. Full consideration of safety attitudes in selection, promotion, and exclusion processes.

Secondary reinforcement and stabilization mechanisms (other cultural artefacts)

7. Organizational structure and reporting lines for the safety function
8. Integration of safety into procedures and management systems
9. Rites and rituals for safety and incorporating safety
10. Site design and layout
11. Stories about key people and events in safety
12. Charters, creeds, and other formal statements

Consultation and participation

13. Consultation of workers and stakeholders before taking decisions affecting their safety
14. Involvement of workers and relevant stakeholders in key safety management activities.

TRUST, PSYCHOLOGICAL SAFETY, AND LEARNING

Trust is a value in itself, with associated beliefs, on which consultation and participation practices and the related artefacts we have just presented are based. This value is particularly important for prevention in work environments characterized by VUCA, where no individual can claim to know all the risks perfectly, let alone all the countermeasures.

Trust is also essential to the shared vigilance we mentioned earlier, and it is particularly important when it comes to managing psychosocial risks.

FOCUS ON... THE ETHICS OF TRUST

Trust is often viewed from a psychological perspective. However, French philosopher Paul Ricoeur affirmed that "trust is not based on psychological factors but on ethical factors. Trust is fundamentally linked to the effectiveness of congruence over time between the word given and the behaviour that follows it. Trust is about keeping promises".

Within organizations, he specified that "in the world of work and of intersubjective relations within the collective, this promise concerns the fairness of judgements made by others about the conduct of *ego*, in the dynamic triangle of work" (the triangle *ego–others–reality* that we presented earlier) (Ricoeur, 1990).

"These judgements cover at least two points:

- Judgement on the practical difficulties actually encountered by ego in the exercise of her work, i.e. in the face of what reality brings out as resistance and failure of the technique in the face of the task.
- Judgement on the quality of the fittings, adjustments, innovations, and discoveries produced thanks to ego's ingenuity".

We can link this with the two types of judgement that Dejours distinguishes in work psychodynamics (Dejours, 1995):

- Utility judgements (technical, economic, or social)
- Conformity judgements

It is when trust and the values and beliefs attached to it (such as fairness and honesty) are embedded in the culture that these judgements become systematically fair, in turn reinforcing the values and beliefs underpinning trust, in the kind of self-perpetuating cultural loop we described earlier.

Developing Trust

To start and boost this virtuous circle of developing trust, Edgar Schein invites leaders to "personize" relationships, which means not just "personalizing" them, but taking them to what he calls "level 2". This level goes beyond purely professional and transactional relationships and involves recognizing the whole person and not just the role she plays in the organization (Schein and Schein, 2018).

Establishing this kind of relationship would be the characteristic mission of what he calls *humble leadership*, which would be all the more imperative the more work requires *cooperation* between workers, in the sense given previously. Of course, the development of these relationships cannot be decreed and does not depend solely on the goodwill of the leader: this requires repeated and sustained efforts to open up, especially in macro-cultures where this kind of relationship is not usual in the professional sphere.

Schein prefers the concept of *openness* to *transparency*, to emphasize that the communication that the leader must put in place (including decisions on what is communicated and how it is communicated) is not a passive process consisting solely of making information visible, available, and easily accessible to others, but a proactive process involving dialogue in which the leader is sincerely committed.

We see this approach as the process that enables workers and leaders to reveal their values and beliefs and discover those of the other, thus enabling each to better understand the other's behaviour and objectives, avoid misunderstandings, anticipate possible discrepancies, and manage them better.

FOCUS ON... TRUST, COOPERATION, AND COMMUNICATION

Stephen Covey described the quality of this communication as a continuum based on trust and cooperation - these two factors being directly and proportionally linked to each other - and in which three levels of communication can be distinguished (Covey, 1989) (Figure 85):

- When trust is low, cooperation is equally low and communication is *defensive*: each party seeks to retain its margins of uncertainty to preserve its autonomy, as we saw above, in a win-or-lose logic.
- When trust develops, so does cooperation (the two feed off each other) and communication becomes *respectful*: political games are still played, but they are so without hostility as everyone accepts that certain concessions will be necessary to maintain good relationships between the parties.
- It is when trust is well established that cooperation is fully expressed and organizations realize their full potential: communication at this stage is *synergistic*. Everyone proactively expresses their perceptions, beliefs, and values, as well as their interests, fears, and weaknesses. Not only are they prepared to listen to those of others, but they actively seek to get to know and understand them to establish win-win cooperative relationships.

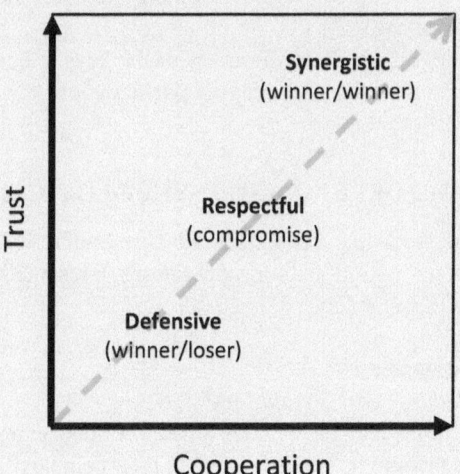

FIGURE 85   Communication levels. (Adapted from S. Covey (1989))

It is in these conditions of trust that the players can truly engage in dialogue on safety, a process in which, to use Peter Senge's expression, everyone *suspends* their values and beliefs in this area, not in the sense of putting them aside, but like a cloth that is laid out for everyone – themselves and others – to examine (Senge, 2006).

Psychological Safety

The concept of **psychological safety** was first used by Edgar Schein and Warren Bennis in the 1960s (Schein and Bennis, 1965), and it came back into the spotlight in the 1990s with the work of William Kahn and, more recently, Timothy Clark and Amy Edmonson (Edmonson, 2019). It can be described as the belief that one will not suffer negative consequences for putting forward ideas contrary to the established consensus, making remarks, voicing concerns, asking questions, or making mistakes.

This concept is close to that of trust, presented above, but they are not entirely identical: trust is used to describe the relationship between two individuals, whereas psychological safety refers to the "climate" within an organization.

We all learn very quickly to regulate our exchanges in our social relationships (including at work) by minimizing interpersonal risks and avoiding showing incompetence or a vision that might be perceived as out of step with that of the group. This sometimes leads us not to ask questions when we have doubts, not to make suggestions even though we may think we came up with valid ideas, or not to defend our point of view as strongly as we may want. Aversion to interpersonal loss (certain and immediate) outweighs the risk of accident (uncertain and remote).

The lack of psychological safety, and the resulting absence of critical feedback and suggestions, may give managers and executives the illusion of safety, but it prevents latent errors from being corrected and improvements and innovations from emerging, thereby increasing the risk of accidents (for a striking example see the Chernobyl case study above).

Edmonson provides an excellent summary of the keys to developing this concept within an organization and thus encouraging participation.

FOCUS ON... DEVELOPING PSYCHOLOGICAL SAFETY

We list below the steps that Edmonson advises leaders to follow, to develop psychological safety within their organizations, presented with a focus on occupational health and safety:

1. **Setting the scene:**
   1.1. *Establishing work's framework*
      - Examine with the workers (operators, managers, engineers, etc.) the tasks to be carried out in all their complexity, especially the interdependencies, and remove ambiguities as much as possible.
      - Ritualize the execution of risk analyses, always with the front-line workers, with a particular focus on identifying uncertainties. Communicate constantly on these aspects of the work.

- Establish opportunities for exchange and dialogue to continuously assess the shared understanding of risk and uncertainty.
- Systematically provide feedback on incidents that have occurred, encouraging learning rather than guilt-tripping.

1.2. *Focusing on purpose*
- Promoting the value of safety: clearly articulating why and for whom prevention is important.
- Constantly communicating (and not just informing) about the issues at stake.

2. **Inviting participation**

2.1. *Demonstrate situational humility*
- Do not try to have all the answers to all the questions.
- Emphasize that we never stop learning and that everyone needs to be humble and curious.

2.2. *Be proactively curious*
- Ask questions with a genuine interest in the answers before expressing your own perspective.
- Demonstrate with your questions that you are curious and interested in detail.

2.3. *Systems and structures*
- Create the structures and tools to bring concerns and ideas to the surface.
- Constantly check that these structures promote a mindset that facilitates open dialogue.

3. **Responding productively**

3.1. *Expressing appreciation*
- Practise active listening and show interest.
- Thank those who raise questions and emit suggestions.

3.2. *Destigmatize mistakes*
- Understand human and organizational performance factors to identify the root causes of accidents.
- Acknowledge feedback on incidents and mistakes to encourage transparency and learn from them.

3.3. *Sanction clear violations*
- Establish clear rules and communicate their importance.
- Respond to violations in a decisive and coherent way.

Learning Organizations

In his reference book *The Fifth Discipline*, which we have already mentioned, Peter Senge defines **learning organizations** as those "where individuals continually develop their capacity to produce the results they want, where new and expansive ways of thinking are fostered, where collective aspiration is unleashed, and where people continually learn to learn together" (Senge, 2006).

In a working environment characterized by the VUCA we have referred to on multiple occasions, made up of intractable systems in which activities and risks are constantly evolving and where even what has been regulated must be continuously reconsidered, it is essential to bring to the organizational level this capacity for learning that we have seen at the individual level.

Although we have followed a different path, we have already dealt with the principles relating to the first four disciplines proposed by Senge in the previous sections (systemic approach, self-control, mental models, and shared vision). The fifth and final discipline, team learning, cannot be developed without first establishing the trust and psychological safety we have just described.

Key Learning Points for the Safety Leader

The **trust** of others (operators, managers, executives, etc.) is essential if the safety leader is to succeed in the mission of effectively guiding their behaviour in the direction of prevention by deploying safety related artefacts and influencing their values and beliefs.

In a climate of **psychological safety**, each worker can count on the benevolence of others to share her views on the risks and her techniques, tactics, and strategies for managing them, to express fears and discontent, and to contribute ideas that will help to raise safety levels and avoid accidents.

Developing and maintaining trust and psychological safety within the organization, which are essential to organizational learning, is far too important an issue to improvise or simply wait for them to arise spontaneously with a show of goodwill.

This approach must inevitably involve some thought as to the objectives and means, and therefore a degree of strategy on the part of the leader, but it is never a question of manipulation. If these efforts to generate trust are to succeed, they must be sincere: the risk is too great if this is not the case, because an approach perceived as deceptive would mean the death of trust and the failure of our efforts to develop a culture of safety.

LEADERS AS GUIDES AND DRIVERS OF CHANGE

We saw in the last chapter that safety leaders are not just visionaries who explain the benefits of prevention to others: it is through their actions, the techniques that they develop, and the practices that they deploy that safety leaders can hope to influence other people's behaviour. To be successful in this role, they need to have an informed understanding of the mechanisms underlying these behaviours, the motivational forces behind people's work, and how they fit together within an organization. But they must also define a strategy, manage it, and drive it forward on a day-to-day basis.

In this final section, we present a few additional keys that the safety leader must take into account to successfully carry out the transformation required to place prevention at the heart of the organization's culture. Here, in the light of our cultural analysis, are some of the classic steps proposed by John Kotter for successfully leading change processes within organizations (if only because, as a reference to this domain, safety leaders are no doubt already familiar with them), sometimes drawing parallels with other approaches and, above all, adapting them to our own prevention objectives (Kotter, 1996).

This is not intended as a detailed method, but rather as a non-exhaustive list of tips so that everyone can define their own strategy and establish their own programme using the concepts and tools set out throughout this book (Blanco-Munoz, 2019).

Take Inventory and Establish the Vision

Identify the Issues and Establish the Urgency

The first step is to determine with the representatives of the prevention function within the organization what are its main health and safety issues. This phase is doubly decisive: firstly, it will be all the easier to mobilize the concerned stakeholders if the issues are well defined; secondly, it will allow for the identification of target techniques and practices to be promoted or, on the contrary, eradicated, and we can begin to examine the values and beliefs (and other practices, techniques, and artefacts) that underpin them.

Focusing on the most critical safety issues, in particular vital risks, enables efforts to be concentrated and action to be prioritized on these risks (which is logical in itself). It also allows leadership to leverage on these major risks to cultivate the values, beliefs, and practices that can then be used to mitigate other less apparent risks.

This is the principle put forward by Tom Krause (Krause, 2015), who recommends that we stop stubbornly attacking the base of Bird's famous pyramid (or Heinrich's, whatever...) in the hope of eradicating the serious events at the top, and instead proceed in the opposite direction: target the vital risks associated with serious accidents at the top of the pyramid and focus efforts on them, to kick-start and give initial impetus to a safety culture that will also lead to a reduction in the frequency of minor incidents.

These two visions are representative of the two typical situations that bring organizations to ask themselves the question of the maturity of their safety culture and the means to advance it: either a serious event or series of events (focus on the top of the pyramid); or the perception on the part of leaders that the organization has reached a "plateau" in terms of the frequency of incidents and therefore needs to use a different approach to overcome it (focus on the base of the pyramid).

The first situation, even if unfortunate in absolute terms, represents a favourable window of opportunity, as safety leaders will be able to surf on the sense of urgency derived from the emotional impact of the triggering events, which is essential for initiating any top-down change process within an organization. In the second case, it would be up to the safety leader to instil this sense of urgency by highlighting the unbearable discrepancy between the overall organization's safety performance and its values and beliefs.

Sometimes it is not possible (or desirable) to establish that famous sense of urgency when it comes to occupational safety: we can think of tense social contexts in which the different stakeholders are not necessarily in a constructive frame of mind and adopt political positions as described above. The safety leader can then adopt the *viral change* method described by Leandro Herrero. This is more subtle and informal, and may seem less heroic than the organizational big bang favoured by Kotter, but may be more effective in certain contexts.

To make it short, Herrero suggests identifying key people within the organization on the basis of their network of influence, and using them as vectors for change, which would take place on a *bottom-up* basis. As in a living organism, these workers would relay significant impulses in the form of specific behaviours that spread by imitation. At a given moment, a tipping point is reached and these behaviours become generalized and therefore culturally shared practices (Herrero, 2006).

Characterizing the Existing Culture

Knowing where we want to go is great, but if we do not know where we are, we are still lost. From the outset, safety leaders often have at least an intuitive vision of the prevention culture they wish to develop in terms of values, beliefs, and practices, but they must start by establishing what the dominant values, beliefs, and practices are in the organization and in the macro-culture (or macro-cultures if the organization is based in several countries and/or regions) in which it is embedded.

The keys to this characterization were provided in the previous chapters, so we will not return to them here.

Uniting Leadership and Defining the Vision

Once the cultural analysis has been carried out and the prevalent values, beliefs, and practices have been identified, we can begin to formulate the vision of the desired target culture, which must at the same time:

- Respond to identified safety concerns, and
- Build on existing values, beliefs, and practices that are conducive to safe behaviour, and challenge those that are not.

In Kotter's approach, if the safety leader driving the process is not the head of the organization, it is the latter who will be the first target that the safety leader will have to make aware of the urgency of changing the culture. It can be argued that, without the full support of the organization's leadership in the broadest sense, no approach of this type would be successful because, as we have seen, it is the leaders who must demonstrate the importance they attach to prevention so that these values can trickle down from the top to the bottom of the organization and generate full support for the new practices.

Once the leaders at the top of the organization are committed to the cause, they have to make sure of two things:

1. The first is to be able to rely on a "core team" of influential people among their direct reports in key functions. Pre-mortem exercises (Klein, 1998) or seminars can be of great help in identifying these people, bringing them together in the form of a coalition to sponsor the project, and integrating them into the process of articulating the vision.

2. The second is to check that the managers at each level will also adopt the safety leadership role and will therefore act as transmission belts. Unless it is a very small structure, top management's leadership must be effectively relayed to each level. To do this, they must communicate, empower, and act as explained below.

In Herrero's *viral change* approach, although it is important to ensure that the vision is accepted by the C-suite, it is not necessarily the hierarchy that cascades the change, but the more or less structured network of influencers led by the safety leader who, even though they may have no formal authority, in turn become safety leaders themselves by promoting and transmitting the key defined behaviours.

Communicate, Empower, and Execute

Convincing All Stakeholders

As we have reiterated, the communication needed to change the safety culture cannot be limited to a propaganda exercise. Effective communication depends more on credibility than quantity.

As for quantity, a poster campaign, memos, or a few town hall meetings are not enough. Leaders must take advantage of every opportunity to highlight their vision and the values, beliefs, and practices they want to develop to achieve it.

This means bringing the topic relentlessly forth into their daily routines, in individual interviews, in team performance reviews, in progress reviews of every project, in every meeting, at the water cooler... They will always seek to raise the subject of safety at the slightest opportunity, by talking about incidents that have occurred, actions in progress, difficulties, and any element, even anecdotal, that helps to illustrate the importance of the subject.

But the most important thing is to back up the sincerity of this communication with practical demonstration. Leaders must embody the culture and set an example at all times, providing immediate and certain feedback on every safety-related behaviour they encounter on a daily basis.

Formalizing Practices

Each organization must clearly establish the key roles and responsibilities in terms of safety, and formalize critical working techniques and management practices so that they can be integrated into the routines of each function, department, and individual. For more details on how these mechanisms and tools can be articulated within a management system, we refer once again the reader to the ISO 45001 standard.

However, it is important to remember that in the real life of intractable systems, not everything can be regulated, and that the main argument of the cultural approach to safety lies in succeeding in motivating preventive behaviour even in the absence of specific instructions, constant supervision, and means of constraint.

As for the processes and activities that we do manage to formalize, whether they are structured in the form of a specific management system or embedded in other systems, they must be geared as much to prescribing safety techniques and practices as to helping to anchor the underlying prevention values and beliefs in the organization.

The kind of ownership that we aim to elicit at all levels of an organization cannot be the result of simply drafting a series of procedures but is just as much a matter of developing a culture of prevention, which requires above all the example and daily focus of safety leaders. In this respect, Herrero stresses the primacy of the viral transmission of expected behaviour, which thus becomes a practice before eventually being regulated.

Removing Obstacles

Standards such as ISO 45001 describe the elements that must be present in a management system but do not mention what must be eliminated from an organization. There are many obstacles to the promotion of prevention values, beliefs, and practices, and safety leaders need to identify those that are present in their organization. For example, we could find:

- Organizational structures and processes (operational as well as support, financial, management, etc.) that are fundamentally misaligned with the target safety culture and in which we would try to "incorporate" the new ways of doing things.
- Individual performance assessment, reward, bonus, incentives, and promotion systems that contradict the new vision and/or expected behaviours.
- People, especially those in positions of responsibility, who, because of their deeply held personal values, beliefs, or habits, would not be capable of evolving toward the expected attitudes.

At the beginning of the journey, as long as the change process has not yet reached a significant level of inertia, it is not always possible to tackle all these obstacles, but leaders must identify and remove at least the most important and obvious ones as quickly as possible. This is not only to move the process forward but also to maintain its credibility if it has been announced as a major programme of change according to Kotter's model.

Empower the Workers and Coach New Leaders

Former US Navy nuclear submarine commander L. David Marquet brilliantly reported the "leader-follower" problem he found in the military and that we can find in every organization. This scheme, which worked beautifully for purely manual work, is no longer adequate in modern world organizations employing essentially technologists, as presented earlier. Although it is seductive and psychologically rewarding for the leaders and reassuring for the followers, it may be debilitating for the latter and result in wasted human potential and suboptimal performance.

As a solution, he proposes the "leader–leader" alternative, based on the belief that "we can all be leaders, and in fact, it's best when we all are leaders. Leadership is not some mystical quality that some possess and others do not. [...] The leader–leader model not only achieves great improvements in effectiveness and morale but also makes the organization stronger. Most critically, these improvements are enduring, decoupled from the leader's personality and presence. Leader–leader structures are significantly more resilient, and they do not rely on the designated leader always being right. Furthermore, leader–leader structures spawn additional leaders throughout the organization naturally. It can't be stopped". (Marquet, 2015).

By implementing a series of practices based on trust, psychological safety, and organizational learning as presented above, Marquet empowered his team members and achieved outstanding results. Among those practices:

- Use guiding principles (shared values) for decision criteria.
- Stop just trying to avoid errors (based on the belief, informed by the human factors knowledge, that there will always be errors) and aim for excellence (a shared value) instead.
- Specify goals, not methods, and resist the urge to provide solutions (based on trust).
- Encourage a questioning attitude over blind obedience (based on psychological safety).
- Finally, do not preach and hope for ownership: implement the mechanisms (practices) that actually give ownership.

The Problem of Metrics – Again

The Western managerial mentality requires indicators to set objectives and measure progress. We have already touched on this thorny issue, so we will just reiterate here that an excessive focus on quantitative result indicators can undermine transparency and generate double talk, as well as dehumanize prevention, with a negative impact on the values and beliefs we are trying to develop.

There is a wealth of information that can be used as a basis for assessing the maturity of the prevention culture, which could be presented as leading indicators of prevention:

- In terms of practices, we can evaluate the effectiveness of prevention training, risk assessment, and participation processes, the *quality* of reporting and analysis of incidents and dangerous situations, etc.
- In terms of beliefs and values, as we mentioned earlier, a number of organizations and consultancy firms offer surveys to assess them, which, if carried out periodically, can give an idea of the current safety climate that can be presented as a proxy to safety culture progress.

It is also possible, and even desirable, to place less emphasis on indicators and more on monitoring the implementation of concrete actions that have an impact on prevention, as shown below.

Building on Early Successes

Any truly transformative change process takes time, and the process risks running out of steam if there is no short-term, achievable success to celebrate, allowing us to encourage and reassure on its progress. The difficulty with prevention is that its main success is, by definition, the absence of an event.

Hoping to reach a symbolic milestone of hours worked without an accident is at once insignificant, debatable, and risky: one unfortunate incident should not be considered a symbol of failure if the safety culture as a whole is moving forward.

Safety leaders must create the first successes instead of waiting and crossing their fingers. Given the aforementioned power of storytelling as a technique for transmitting and sharing beliefs and values, they can in particular put in place high-profile flagship programmes, on which they need to invest the necessary resources and focus to make them succeed quickly (for example, improving the ergonomics of a particularly arduous job).

This initial visible and meaningful achievement can then be put forward to enhance the credibility of the approach and nurture the underlying values and beliefs.

Ensuring Sustainability

It Always Takes Longer than Expected

A change of culture takes years to consolidate. Declaring success too soon can be a fatal and lamentable mistake once so much energy has been spent; it can halt the momentum of progress and allow the forces resisting change to regain the upper hand.

The fact that the accident's FR is on a positive trend and that significant battles have been won, as mentioned above, may encourage both safety leaders and those who have not yet embraced the new culture to declare that war is over. The former because they are over-enthusiastic, the latter because they see this as an opportunity (consciously or subconsciously) to stop bothering and return to their old habits.

Anchoring Safety at the Heart of the Organization's Culture

The main advantage of working on the development of a prevention culture is that, by succeeding in having the values, beliefs, techniques, and practices linked to this culture shared by every individual in the organization, safe behaviour is sustainably embedded in them. Newcomers, joining a group where what is acceptable and rewarded is self-evident, adopt these practices and, gradually, spouse the underlying values and beliefs.

But we must remember that if efforts were needed to overcome inertia and transform habits, it was because there were forces at play that had installed them and made them natural, values and beliefs that were justified in their context and that had to be fought. It would be naïve to think that these forces can disappear forever within an organization, if only because its members continue to be influenced by the beliefs, values, and practices of the macro-culture and other "moral circles" to which they also belong. Moreover, all organizations are subject to upheavals that can upset short-term priorities and seemingly justify the acceptance of certain "bad compromises", opening the way to regression.

There are two key ways of consolidating a culture of prevention:

The first is to maintain a high level of communication, ingrained in day-to-day routines, that explicitly links good safety performance to the excellence of the culture that the organization has managed to put in place. Every worker, in every function and at every level, must be aware of the link between their attitudes and behaviour, within the framework of the organization's culture, and its excellent safety record. Safe working techniques and related management practices must be highlighted as a source of pride for the organization and for each worker.

The second is to take steps to ensure that the next generation of leaders at the top of the organization and in each of its departments will truly embody the values and beliefs of the safety culture. One failed succession plan can wipe out a decade of effort. Particularly at the top, the board of directors must understand the dynamics of safety culture and be fully aware of the crucial importance of their choices when it comes to deciding who should take the lead in maintaining and further developing it.

Key Learning Points for the Safety Leader

There can be no safety culture without safety leadership. Safety leaders play a key role in ensuring that these topics are effectively placed at the heart of an organization's culture. As such, they must embody the values and beliefs associated with safety and set an example by demonstrating their unwavering commitment to devising, implementing, monitoring, sponsoring and embodying safety practices.

The sincere adoption of these attitudes and practices is a necessary but not sufficient condition for them to succeed in this mission. They must have a very good knowledge of human factors and the cultural dynamics of organizations, enabling them to formulate a vision and a strategy for change at this level, as well as the specific tools to implement it.

We hope that the material presented in this book will be of use to all safety leaders.

5 Conclusion
Having To, Being Able To, and Doing

"When we want we can, when we can we must."

Napoleon Bonaparte

Having looked at the various factors that influence workers' behaviour in the face of risk, we can see that this is a complex phenomenon, but that we now possess a number of keys that enable us to make progress in our understanding of the subject, as well as tried and tested tools for managing it more effectively.

This is an issue that cannot be approached solely from the point of view of the individual, the community, or the system. It cannot be studied only in terms of management, ergonomics, sociology, psychology, or neuroscience. Nor can we hope to solve it by taking either the rule-based safety or the managed safety side.

By looking at all these different dimensions, we can broaden our horizons and consider new perspectives. And this also allows us to imagine how they can be combined in a holistic way and how we can integrate different tools from a wide range of disciplines into a coherent and effective approach to prevention.

In this final chapter, and as a conclusion to our journey, we propose a summary that we hope will enable the safety leader to recapitulate the main concepts that have been set out and to visualize the logic with which they fit together. We will illustrate this summary with a model that we will call the **three traffic lights**, or **"I should/I can/I do"**.

OUR NATURAL TENDENCIES

To begin with, it should always be borne in mind that the reason why the human factor is often considered to be the cause of the problem is simply because, quite often, adopting safe behaviour represents an effort for the worker and/or the organization and, as a result, unsafe shortcuts are taken (Figure 86).

This effort may be physical or intellectual, involving a greater expenditure of time or energy. Or it may be financial, calling on technological, material, or human resources. But it can also be relational, requiring the worker to negotiate with other players (co-workers or customers, superiors or subordinates, other functions or authorities, etc.) to obtain their acceptance and/or contribution.

Beyond the automatisms and other habits that we have examined and that bring their own challenges, the always imperfect analysis that we will make of the costs and benefits of the actions that we intend to carry out often pushes us towards the easy path. To the point that quite often, as Simon exposed some 70 years ago, such easy path is the only one we really consider.

DOI: 10.1201/9781003617341-5

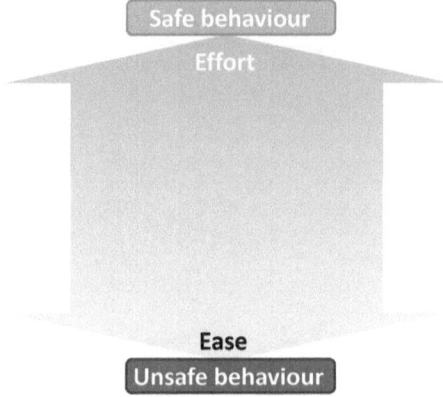

FIGURE 86   Ease vs effort (drawing by author).

We have seen the limits of our perception and those of our rationality (local, bounded, and necessarily subjective, as well as multidimensional and influenced by our emotions). To sum up, there is our tendency to take the shortest route, as well as our inability to assess, in a factual, methodological and objective way, the risks and our capacity to manage them.

If it were otherwise, neither this book nor the rest of the literature devoted to human factors, safety management, or behavioural safety would have any place: if not survival instinct, common sense would be in charge of guiding our every action along the straight and narrow path of safety.

But as former French Prime Minister Jean-Pierre Raffarin once said, even if the path is straight, the slope is steep. Bringing individual and collective action towards safe behaviour therefore involves a series of stages.

I SHOULD

The first step in getting a worker (alone or as part of a team) to adopt a behaviour that is not the easiest at first sight, because it requires extra effort, is to define the expected behaviour. This means establishing unambiguously what individuals should do in a given situation. And there are two dimensions to this work.

THE MANAGERIAL DIMENSION

When it comes to defining expected behaviour, we tend initially (and sometimes exclusively) to think about the rules that must be followed and the roles of each of the players in terms of responsibilities and authority, as well as the notions of hierarchy, supervision, and accountability that go with them. They are indeed important, and we are adding these elements to our diagram illustrating the path leading to the expected behaviour, on the left-hand side in the figure below, the side where we are going to represent the **managerial dimension** (Figure 87).

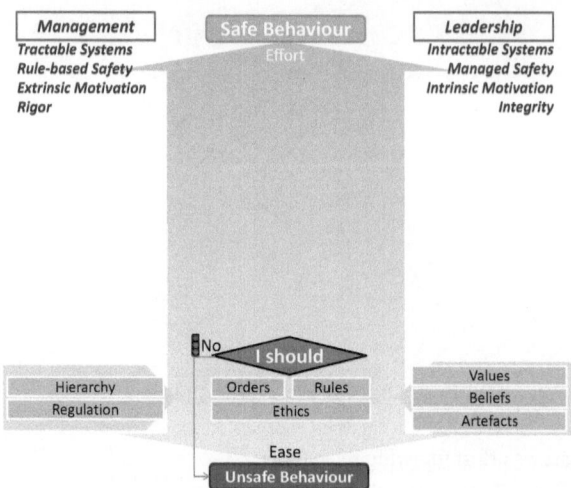

FIGURE 87   First traffic light – I should (drawing by author).

This corresponds to the concepts of rule-based safety and calls for rigour in engineering, planning, execution, and control. As we have seen, it is particularly well-suited to docile or tractable systems (typically characterized by low complexity, lending themselves to tight coupling) but is not very effective once we move beyond what is perfectly known and predictable.

This approach is fully justified and, to some extent, necessary in high-risk industries and activities, but its generalization in other contexts and the tendency to focus exclusively on it are more or less implicitly based on McGregor's Theory X, which is grounded on the belief that workers seek above all to make their lives easier and only make efforts if they are forced to do so by extrinsic motivating factors (rewards and punishments).

THE LEADERSHIP DIMENSION

However, we have seen that this vision of work is truncated and incomplete. Workers at all levels can also be motivated by intrinsic factors such as a sense of belonging, autonomy, development, pride in a job well done, etc. (Theory Y), which are based on values and beliefs rooted in them, which form the foundations of their work ethic and with which they will align their behaviour out of integrity and not out of obligation.

Anchored in their emotions – the neuropsychological underpinnings of which we are beginning to understand – such intrinsic motivation will guide their behaviour in a way that is more subtle but often more effective than coercion, especially in intractable systems where not everything can be regulated and controlled. They adopt these values and beliefs by evolving within a culture, which can, in turn, be shaped by leaders through their daily engagement and the deployment of various artefacts.

We add these other elements to the right-hand side of the diagram, where we represent the **leadership dimension**.

At this first level of defining behavioural expectations, we can already see an initial complementarity between the managerial and the leadership approaches:

- The former is undoubtedly more prescriptive, Cartesian, and structured, and may therefore seem more reassuring for both management (which thus fulfils its traditional role) and operators (who can tell themselves that as long as they follow the instructions, if there is any problem it will not be their fault).
- The second is more ethereal and difficult to build and consolidate, but it imposes itself on workers in all circumstances, even when no one is watching.

We can understand that if this first stage is not completed by working in a balanced way on both levels (i.e. by putting more or less emphasis on one or the other depending on the complexity and dangerousness of each task), it is impossible for workers to act in the way we would like them to. We would be stuck in this "first red light".

But just having properly defined the expected behaviour – both in terms of rules and in terms of what we have called a "safety ethic" – does not mean that everyone will comply...

I CAN

Once the first step has been taken, the expected behaviour needs to be made easier: it needs to be feasible and as unrestrictive as possible compared with alternative unsafe behaviours. Is not just that an easier alternative will be preferred following a comparative evaluation: safer options may very well be subconsciously discarded without being really considered.

THE MANAGERIAL DIMENSION

On the managerial side, there is plenty to do. This starts at the design and engineering stage, with the definition and implementation of equipment and working methods that reduce the level of risk at its source as far as possible and provide the means to control residual risks. Tools such as fault trees, FMECA, and HAZOP should be used to identify potential accidental events, as well as ergonomic problems at workstations and in human–machine interfaces.

The better we succeed in integrating risk management at this level right from the design stage, the less we will need to add safety rules to procedures, instructions, and other work standards. And that means consulting and involving all the workers concerned, who are confronted with the job "reality" on a daily basis, particularly operators and those who will be responsible for maintaining the equipment.

Next, it is necessary to identify and ensure the level of competence, skill, and dexterity required by the personnel carrying out the task in question. It will be necessary to select the most suitable individuals in terms of physical, intellectual, and psychological fit, experience and expertise... and/or to develop the required qualities through appropriate training programmes.

Thirdly, we need to be sure that the resource allocation processes in place allow us to assign the material and human resources needed to carry out the activity in the required safety conditions. This is the primary role of management in the traditional sense of the term.

Finally, the work environment must be controlled to prevent factors external to the job and the activity in question from interfering with it and endangering those involved, either directly or through distractions and other error precursors that we have studied.

THE LEADERSHIP DIMENSION

But the leadership dimension is just as important at this stage. All the managerial activities listed above will only be prioritized and carried out to the extent that safety has been integrated by everyone, at every level and in every function of the organization, as a central and essential value weighing on choices and trade-offs, whether they are strategic, tactical, or technical; financial or organizational; highly dimensional or routine.

Safety leaders must also invest in coaching other workers, not only to develop their skills and know-how but also to actively promote safety attitudes and, as far as possible, encourage other safety leaders to emerge.

In addition to all this, they must ensure that they establish and support a climate of psychological safety in which every worker can feel completely at ease to raise fears, ask questions, make comments, or suggest ideas for improvement, thus creating one of the essential conditions for organizational learning, leading to better control and a reduction in the level of risk. This is the *sine qua non* for the organization to benefit fully from the vigilance, commitment, and expertise of each individual, and for each individual to benefit from that of others (Figure 88).

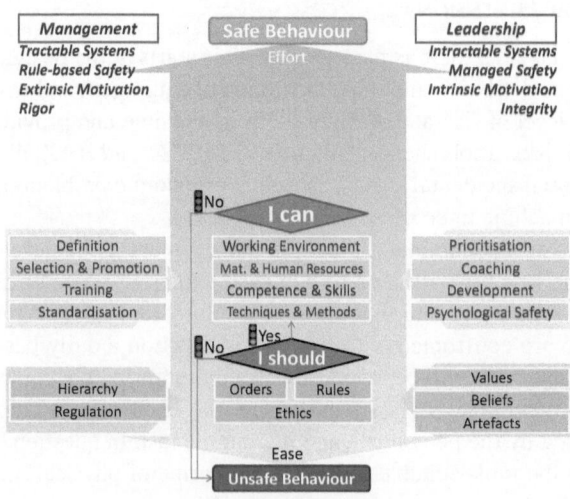

FIGURE 88   Second traffic light – I can (drawing by author).

If we fail to make all these elements available and combine them appropriately ("second red light"), not only we are back to square one, because we are not making it easy to carry out the work as it has been defined so that it can be done safely, and more generally, the preventive behaviours as they are expected: we are also generating frustration among workers who would know what they need to do to avoid accidents and would be aware of the importance of safety, but who would not be able to act accordingly.

But having successfully passed this second traffic light is no guarantee that workers will systematically adopt the expected behaviours...

I DO

Indeed, there is a third and final step to negotiate: we need to consolidate and perpetuate the expected behaviours across the board. We need to ensure that everyone knows what to do in a given situation, has the means to act accordingly, and always makes the extra effort expected of them.

Here we are faced with the challenge of day-to-day work, the pull for efficiencies, successive emergencies and crises, changing priorities, and the constant search for the arrangements needed to keep an organization running in all its complexity.

THE MANAGERIAL DIMENSION

In the managerial dimension, we have to exercise effective supervision – or as far as it is possible for supervisors and acceptable to supervisees – to ensure compliance with rules and instructions, which leads to its share of debates and even arguments because, once again, no rule will ever be perfectly adapted to 100% of imaginable cases (and even less to those we cannot imagine). Disciplinary schemes will have to be implemented to manage the violations, but adjustments and compromises will sometimes be necessary.

THE LEADERSHIP DIMENSION

This brings us back to the question of leadership. Safety leaders not only accompany and ensure, by providing immediate and constant feedback, that prevention techniques and practices are systematically deployed and remain in force: they lead by example, walk the talk, and are personally engaged over the long term, earning the trust of their counterparts.

The climate of psychological safety that they have succeeded in establishing means that they can count on open and constructive dialogue between all stakeholders to ensure that these rules are discussed, reviewed, and updated as necessary – which is essential to guarantee everybody's buy-in – always in the light of the central importance that everyone attaches to safety. As a result, the management compromises mentioned above will therefore be the "right compromises".

Leaders encourage every worker to adopt prevention techniques and practices, which become integral artefacts of the organization's culture. All this is not just talk but a reality made up of actions, large and small, that are continuously perceived by

FIGURE 89   Third traffic light – I do (drawing by author).

each worker, who becomes imbued with the underlying values and beliefs that promote safety and, in turn, becomes an active vector of these values, influencing their colleagues and, more broadly, all those with whom they interact.

Hazards are better perceived and draw workers' attention. Safe behaviour turns into automatisms, and safe practices become habits. Both spontaneous and deliberate behaviour is now geared towards prevention and safety, values that infuse the conscious and subconscious mind with an emotion capable of triggering safe behaviour and inhibiting unsafe behaviour. Shared vigilance is no longer a slogan but a common practice.

It is by working on this last stage that we can hope to pass this third and final "traffic light", thus firmly and durably anchoring safety at the heart of the organization's culture and in that of each worker (Figure 89).

A FINAL WORD

Now that we have the full picture, we can realize that there is no point in opposing the regulated approach against the managed approach to safety, "Safety I" against "Safety II" (or "Safety Differently"), behavioural safety against safety culture or values-based safety, just as one might try to pit one type of leadership against another or justify the pre-eminence of technology as a lever for progress in safety against that of human factors.

For most of his long and prolific career, the oft-quoted management guru Peter Drucker remained deliberately ambiguous about the notion of leadership, a subject on which he did not like to express himself as such and on which he did not write a single paper. For him, leadership was an integral part of the mission and work of management. It was not that he did not think the two concepts could be distinguished: he just did not want them to be seen as separable, even less opposed in any way.

We need to adopt the same philosophy when it comes to safety and overcome the dichotomies that have arisen between different "schools", undoubtedly fuelled by a certain polarization bias. Because in most cases and for the most part, not only are these two management/leadership visions (and the different contributions that flow from them) not antinomic or even just complementary: they are both necessary and, moreover, synergistic, reinforcing each other.

The tools, techniques, and practices deployed as part of safety management systems are all the more effective if they are deployed within an organization where safety is a core value and, in turn, consolidate the importance of safety by becoming cultural artefacts. Just as the work of safety leaders in guiding teams to adopt preventive behaviour can only be facilitated by the presence of rules, processes, and systems for managing safety (or by concomitant efforts to put them in place).

After a century of pushing Taylorism to the limit, many companies are now exploring less vertical, less siloed, and more participative models, such as autonomous production units and quality circles, which give more meaning to everyone's work and enable them to truly embrace values such as excellence, service, innovation... and safety. There is no doubt that these and similar models, which simultaneously promote rigour and trust, enable the various dimensions of safety to be combined more effectively.

As this author has often put it, this is neither idealism nor ideology but the best way to approach the topic of behaviour: treating humans as they are and not as they are supposed to be (Blanco-Munoz and Pater, 2024).

We hope that the vision and tools presented in this book will help safety leaders, whatever the model of their organization, to make informed and relevant contributions to moving forward in this journey.

Bibliography

Adams J. (1981).The efficacy of seat-belt legislation: A comparative study of road accident fatality statistics from 18 countries. University College London. Accessed from https://trid.trb.org/view/171184

Ade C.J., Turpin V.G., Parr S.K. et al. (2021). Does wearing a facemask decrease arterial blood oxygenation and impair exercise tolerance? *Respiratory Physiology & Neurobiology*, 294, 103765.

Adelson E.H. (2005). *Checkershadow Illusion: Perceptual Science Group*. MIT. Accessed from https://persci.mit.edu/gallery/checkershadow//

Adler A. (1938). *Social Interest: A Challenge to Mankind*. J. Linton and R. Vaughan (Trans.). Faber and Faber Ltd.

Ajzen I. and Fishbein M. (2005).The influence of attitudes on behavior. In: Albarracín D., Johnson B.T., Zanna M.P. (Eds.) *The Handbook of Attitudes* (pp. 173–221). Erlbaum.

Allen J.G., Mac Naughton P., Satish U., Santanam S., Vallarino J. and Spengler J.D. (2015). Associations of cognitive function scores with carbon dioxide, ventilation and volatile organic compound exposures in office workers: A controlled exposure study of green and conventional office environments. *Environmental Health Perspectives*, 124(6), 805–812.

Alpert M. and Raiffa H. (1982). A progress report on the training of probability assessors. In: Kahneman D., Slovic P. and Tversky A. (Eds.). *Judgment under Uncertainty: Heuristics and Biases* (pp. 294–305). Cambridge University Press.

Alvesson M. and Spicer A. (2016). *The Stupidity Paradox*. Profile Books.

Amalberti R. (2001).The paradoxes of almost total safe transportation systems. *Safety Science*, 37 (1–2), 109–126.

Anderson B.L. (2003). Perceptual organization and White's illusion. *Perception*, 32, 269–284.

Ariely D. (2010). *Predictably Irrational* (revised ed.). HarperCollins.

Asch S. (1987). *Social Psychology*. Oxford University Press.

Atkinson R. and Shiffrin R. (1968). Human memory: A proposed system and its control processes. In: Spence K. (Ed.) *The Psychology of Learning and Motivation: Advances in Research and Theory* (pp. 89–195). Academic Press.

Baddeley A. and Hitch G. (1974). Working memory. In: Bower G. (Ed.) *The Psychology of Learning and Motivation* (pp. 48–79). Academic Press.

Baddeley A. (2000). The episodic buffer: A new component of working memory? *Trends in Cognitive Sciences*, 4(11), 417–423.

Bandura A. (1986). *Social Foundations of thought and Action: A Social Cognitive Theory*. Prentice-Hall.

Bandura A. (1999). Social cognitive theory: An agentic perspective. *Asian Journal of Social Psychology*, 2. 21–41.

Bar-Yam Y. (2002). General Features of Complex Systems. In: Knowledge Management, Organizational Intelligence and Learning, and Complexity – Vol. I. UNESCO – Encyclopedia of Life Support Systems (EOLSS).

Birch J. (2012). Worldwide prevalence of red-green color deficiency. *Journal of the Optical Society of America A-Optics Image Science and Vision*, 29(3), 313–320.

Blanco-Munoz E. (2016). Can Indicators Kill? 6 Steps to Murdering with Figures. LinkedIn. Accessed from https://www.linkedin.com/pulse/can-indicators-kill-6-steps-murdering-figures-eduardo-blanco-munoz/

Blanco-Munoz E. (2016). Six Keys to Setting Safety Rules… so people actually follow them. LinkedIn. https://www.linkedin.com/pulse/6-rules-making-every-manager-should-know-eduardo-blanc o-munoz/?trackingId=sEnxd7sfRFG3jCnm9J4GSA%3D%3D

Blanco-Munoz E. (2019). Culture de la prévention: théorie et pratique des comportements face aux risques. Collection Techniques de l'ingénieur, réf. SE3815.

Blanco-Munoz E. (2020). Analyse des événements et enquêtes post-accidents en santé et sécurité au travail. Collection Techniques de l'ingénieur réf SE3944.

Blanco-Munoz E. (2021). Erreurs: causes, conséquences et modes de gestion en santé et sécurité au travail. Collection Techniques de l'ingénieur, réf. SE3948.

Blanco-Munoz E. and Pater R. (2024). Safety change mastery: A transcontinental conversation on six best strategies and practices. *Professional Safety*, 69(9), 18–23.

Blake R. and Mouton J. (1964) The Managerial Grid: Problems and Possibilities of Improving Production Through Participation of Peaople. Gulf Publishing.

Bloch A. (1978). *Murphy's Law and Other Reasons Why Things Go Wrong*. Price/Stern/Sloan Publishers, Inc.

Bloom B.S. (Ed.) (1985). *Developing Talent in Young People*. Ballantine Books.

Bouchard A.E., Dumas E. and Fecteau S. (2025). Attention and decision making. In: Grafman J.H. (Ed.) *Encyclopedia of the Human Brain* (pp. 316–332, 2nd Edition). Elsevier.

Bradford D. and Robin C. (2021). *Connect: Building Exceptional Relationships with Family, Friends and Colleagues*. Currency.

Broadbent D. (1958). *Perception and Communication*. Pergamon Press.

Burchman T.J. and Miller E.K. (2007). Top-down versus bottom-up control of attention in the prefrontal and posterior parietal cortices. *Science*, 315, 1860–1862.

Camerer C., Loewenstein G. and Weber M. (1989). The curse of knowledge in economic settings: An experimental analysis. *Journal of Political Science*, 97(5), 1232–1254.

Carlyle T. (1888). *On Heroes, Hero-Worship and the Heroic in History*. Fredrick A. Stokes & Brother.

CDC (Center for Disease Control). (2024). Understanding the Opioid Overdose Epidemic. Accessed from https://www.cdc.gov/overdose-prevention/about/understanding-the-opioid-overdose-epidemic.html on Apr 5, 2025.

Chiaramonte C. (2007). Attraction de l'attention par des différences relatives de familiarité : étude expérimentale de la dynamique des processus précoces. Thèse de doctorat en Psychologie cognitive, soutenue à Grenoble 2. Accessed from https://www.theses.fr/2007GRE29034

Clerc D. and Piriou J.P. (2011). *Lexique de sciences économiques et sociales (9ème édition)*. La Découverte.

Colrain I.M., Nicholas C.L. and Baker F.C. (2014). Alcohol and the sleeping brain. *Handbook of Clinical Neurology*, 125, 415–431.

Cooper D. (2000). Towards a model of safety culture. *Safety Science*, 36, 111–136.

Cooper D. (2001). *Improving Safety Culture: A Practical Guide*. Applied Behavioural Sciences.

Corson Y. and Verrier N. (2013). *Les faux souvenirs*. De Boeck Supérieur.

Covey S.R. *The 7 Habits of Highly Effective People*. Simon and Schuster (1st ed. 1989, 25th anniversary ed. 2004).

Crozier (M.) and Friedberg E. *L'acteur et le système*. Éditions du Seuil (1977, réimpression d'avril 2014).

Dab W. (2018). *La santé et le travail (2nde édition)*. Arnaud Franel.

Dahlhamer J.M., Connor E.M., Bose J., Lucas J.W. and Zelaya C.E., (2019). Prescription Opioid Use Among Adults with Chronic Pain: United States. National Health Statistics Reports, No.162, August (2021).

Damasio A.R. (1994). *Descartes' Error: Emotion, Reason and the Human Brain*. A. Grosset/Putnam Books.

Damasio A.R. (2003). *Looking for Spinoza: Joy, Sorrow, and the Feeling Brain*. Mariner Books.

Daniellou F.S., Simard M. and Boisseres I. (2010). *Facteurs humains et organisationnels de la sécurité industrielle: un état de l'art*. Fondation pour une Culture de Sécurité Industrielle. Accessed from https://www.foncsi.org/fr/cahiers/

Dejours C. *Le facteur Humain*. Presses Universitaires de France/Humensis (1995, 8ème édition 2022).

Dekker S. (2014). *The Field Guide to Understanding Human Error*. Ashgate.

Dekker S. (2014). The bureaucratization of safety. *Safety Science*, 70, 348–357.

Deming W.E. (1986). *Out of the Crisis* (2nd printing ed.). MIT Center for Advanced Engineering Study.

Dijksterhuis A. (2007). *Het slimme onbewuste. Denken met gevoel*. Uitgeverij Bert Bakker.

Dore F. and Mercier P. (1992). *Les fondements de l'apprentissage et de la cognition*. Presses Universitaires du Septentrion.

Dortier J.-F. (2011). Le mythe des trois cerveaux. *Sciences Humaines*, 14, 14–17. Accessed from https://lecerveau.mcgill.ca/flash/capsules/articles_pdf/trois_cerveaux_dortier.pdf

Drucker P.F. (1966). *The Effective Executive*. Harper & Row Publishers.

Drucker P.F. *Management: Tasks, Responsibilities, Practices*. Harper & Row (1st ed. 1974, revised ed. 2008).

Duckworth A. (2016). *Grit: Why Passion and Resilience are the Secrets to Success*. Vermilion.

Duhigg C. (2012). *The Power of Habit: Why We Do What We Do and How to Change*. Random House.

Dunbar R.I.M. (1992). Neocortex size as a constraint on group size in primates. *Journal of Human Evolution*, 22(6), 469–493.

Dupré M. and Le-Coze J.-C. Une approche ethnographique de la sécurité industrielle dans la chimie. *Séminaire du GRETS, GRETS* (Groupe de Recherche Energie Technologie et Société), Mars 2023, Paris, France.

Duval C., Piolino P., Bejanin A. and Laisney M. (2011). La théorie de l'esprit: aspects conceptuels, évaluation et effets de l'âge. *Revue de neuropsychologie*, 3, 1.

EDF. Le pré-job briefing. Accessed from https://lesbonsgestesedf.fr/faq/jobBriefing.html on April 10, 2025

Edmonson A.C. (2019). *The Fearless Organization: Creating Psychological Safety in the Workplace for Learning, Innovation, and Growth*. Wiley.

Ekman P. (1999). *Handbook of Cognition and Emotion*, John Wiley & Sons.

Emery F.E. and Trist E.L. (1965). The causal texture of organizational environments. *Human Relations*, 18(1), 21–32.

Endsley M.R. (1995). Toward a theory of situation awareness in dynamic systems. *Human Factors*, 37(1), 32–64.

Epstein M.L., Lazarus A.D., Calvano T.B., Matthews K.A., Hendel R.A., Epstein B.B. and Brosvic G.M. (2002). Immediate feedback assessment technique promotes learning and corrects inaccurate first responses. *The Psychological Record*, 52, 187–201.

ESA ACT (Australian Capital Territory Emergency Services Agency). Dynamic Risk Assessment Overview. Accessed from https://esa.act.gov.au/sites/default/files/wp-content/uploads/dynamic-risk-assessment-overview.pdf on April 10, 2025.

EU-OSHA (Agence européenne pour la sécurité et la santé au travail). *Participation des travailleurs au processus de promotion de la santé et de la sécurité au travail – Guide pratique*. Office des Publications de l'Union Européenne (2012).

Eurocontrol. *Systems Thinking for Safety: Ten Principles*. European Organization for the Safety of Air Navigation (2014).

Fanning R. and Gaba D. (2007). The role of debriefing in simulation-based learning. *Simulation in Healthcare*, 2(2), 115–125.

Fazio R. (1990). Multiple processes by which attitudes guide behavior: The MODE model as an integrative framework. *Advances in Experimental Social Psychology*, 23, 75–109.

Fédération Française du Bâtiment (FFB). Lutte contre les addictions: les actions à mettre en place ! Focus No. 147, May (2024).

Festinger L. (1985). *A Theory of Cognitive Dissonance*. Row, Peterson and Company (1957, 2nd ed). Stanford University Press.

Fiedler F.E. (1964). A contingency model of leadership effectiveness. In: Berkowitz L. (Ed.) *Advances in Experimental Social Psychology* (vol. 1, pp. 149–190). Academic Press.

Fino E. *Transmission et plasticité activité-dépendante au niveau des synapses cortico-striatales*. Neurosciences. Université Pierre et Marie Curie – Paris VI (2007). Accessed from https://theses.hal.science/tel-00811483 on April 10, 2025.

Fishbein M. and Ajzen I. (1975). *Belief, Attitude, Intention and Behaviour: An Introduction to Theory and Research*. Addison-Wesley.

Fleming M. Developing a draft safety culture maturity model. HSE Books (2000).

Flin R., Mearns K., O'connor P. et al. (2000). Measuring safety climate: Identifying the common features. *Safety Science*, 34 (1–3), 177–192.

Flyvbjerg B. and Gardner B. (2023). *How Big Things Get Done, The Surprising Factors Behind Every Successful Project, from Home Renovations to Space Innovation*. Currency.

Fricke T.R., Tahhan N., Resnikoff S., Papas E., Burnett A., Ho S.M., Naduvilath T. and Naidoo K.S. (2018). Global prevalence of presbyopia and vision impairment from uncorrected presbyopia. *Ophthalmology*, 125(10), 1492–1499.

Fritz J., VoPham T., Wright K.P. Jr. and Vetter C. (2020). A chronobiological evaluation of the acute effects of daylight saving time on traffic accident risk. *Current Biology*, 30(4), 729–735.

GALLUP. What Percentage of Americans Smoke Marijuana? (2024). Accessed from https://news.gallup.com/poll/284135/percentage-americans-smoke-marijuana.aspx on April 10, 2025.

Gershon M.D. (1999). *The Second Brain: A Groundbreaking New Understanding of Nervous Disorders of the Stomach and Intestine*. Harper Perennial.

Gillespie R. (1991). *Manufacturing Knowledge: A History of the Hawthorne Experiments*. Cambridge University Press.

Gneezy U. and Rustichini A. (2000). A fine is a price. *Journal of Legal Studies*, 29(1), 468061.

Goldstein B.E. (2011). *Cognitive Psychology: Connecting Mind, Research, and Everyday Experience*. Cengage Learning.

Goszczynska M. and Roslan A. (1989). Self-evaluation of drivers' skill: A cross-cultural comparison. *Accident Analysis & Prevention*, 21(3), 217–24.

Griffiths P. (1998). *What Emotions Really Are: The Problem of Psychological Categories*. University of Chicago Press.

Gulati R. (2022). *Deep Purpose: The Heart and Soul of High-Performance Companies*. Harper Business.

Guldenmund F.W. (2007). The use of questionnaires in safety culture research – an evaluation. *Safety Science*, 45(6), 723–743.

Habermas J. (1987). *Théorie de l'agir communicationnel*. Fayard.

Haines M.R. and Kian D.V.S. (1991). Assessing safety performance after the era of the LTI. In *SPE International Conference and Exhibition on Health, Safety, Environment, and Sustainability* (pp. SPE-23217). SPE.

Hale A. and Borys D. (2013). Working to rule, or working safely? Part 1: A state of the art review. *Safety Science*, 55, 207–221.

Hale A. and Borys D. (2013). Working to rule, or working safely? Part 2: The management of safety rules and procedures. *Safety Science*, 55, 222–231.

Hall E.T. (1959). *The Silent Language*. Doubleday.

Hameiri B., Bar-Tal D. and Halperin E. (2019). Paradoxical thinking interventions: A paradigm for societal change. *Social Issues and Policy Review*, 13(1), 36–62.

Hebb D. (1949). *The Organization of Behavior*. Wiley & Sons.

Heimbeck D., Frese M., Sonnentag S. and Keith N. (2003). Integrating errors into the training process: The function of error management instructions and the role of goal orientation. *Personnel Psychology*, 56, 333–361.

Heinrich H.W. (1931). *Industrial Accidents Prevention: A Scientific Approach.* McGraw-Hill.

Herrero L. (2006). *Viral Change.* Meetingminds.

Hersey P. and Blanchard K. *Management of Organizational Behavior - Utilizing Human Resources.* Prentice Hall (1969).

Herzberg F., Mausner B. and Snyderman B.B. (1959). *The Motivation to Work* (2nd ed.). John Wiley & Sons.

Hofstede G., Hofstede G. and Minkof M. (2010). *Cutures and Organizations: Software of the mind* (3rd ed.). McGraw-Hill.

Holden B.A., Fricke T.R., Wilson D.A., Jong M., Naidoo K.S., Sankaridurg P., Wong T.Y., Naduvilath T.J. and Resnikoff S. (2015). Global prevalence of myopia and high myopia and temporal trends from 2000 through 2050. *Ophthalmology*, 123(5), 1036–1042.

Holland J.H. (1992). Complex adaptive systems. *Daedalus*, 121(1), 17–30.

Holling C.S. (1973). Resilience and stability of ecological systems. *Annual Review of Ecology and Systematics*, 4, 1–23.

Hollnagel E. and Amalberti R. (2001). The Emperor's New Clothes or Whatever Happened To "Human Error"? In *Proceedings of the 4th International Workshop on Human Error, Safety and Systems Development* (pp. 1–18). Sweden: Linköping University.

Hollnagel E. (2009). *The ETTO Principle: Efficiency-Thoroughness Trade-off: Why Things That Go Right Sometimes Go Wrong.* Ashgate.

Hollnagel E. (2014). *Safety-I and Safety-II.* CRC Press.

Homer P.M. and Kahle L.R. (1988). A structural equation test of the value-attitude-behavior hierarchy. *Journal of Personality and Social Psychology*, 54(4), 638–646.

HSE (Health and Safety Executive). Safety Climate Tool. Accessed from https://books.hse.gov.uk/Safety-Climate-Tool

Hudson P.T. (2001). Safety management and safety culture: The long, hard and winding road. In: Pearse W., Callagher C., Bluff L. (Eds.) *Occupational Health and Safety Management Systems: Proceeding of the First National Conference* (pp. 3–32). Crown Content.

IAEA (International Atomic Energy Agency). Safety Culture: A report by the International Nuclear Safety Advisory Group. Safety Series No 5-INSAG-4 (1991).

ICSI (Institut pour une culture de sécurité industrielle). Diagnostique culture de sécurité. Accessed from https://www.safety-in-practice.org/diagnostic-culture-securite

ICSI (Institut pour une culture de sécurité industrielle). L'essentiel du leadership en sécurité (2022). Accessed from https://www.icsi-eu.org/publication/essentiel-leadership-securite

INSERM. Chronobiologie: les 24 heures chrono de l'organisme. Accessed from https://www.inserm.fr/dossier/chronobiologie/#:~:text=L'horloge%20interne%2C%20m%C3%A9tronome%20de%20l'organisme&text=Chez%20l'humain%2C%20cette%20horloge,oscillant%20sur%20environ%2024%20heures on April 10, 2025 (published on June 13, 2017),

ISO 10075-2:1996. Principes ergonomiques relatifs à la charge de travail mentale — Partie 2: Principes de conception.

Johnson S. (2018). *Farsighted.* John Murray.

Jung C.G. (2016). *Psychological Types* (1st ed. 1921). Taylor & Francis.

Kahle L.R. (1983). *Social Values and Social Change: Adaptation to Life in America.* Praeger.

Kahneman D. and Tversky A. (1974). Judgement under uncertainty: Heuristics and biases. *Science*, 185(4157), 1124–1131.

Kahneman D. and Tversky A. (1977). *Intuitive prediction: Biases and corrective procedures.* In: Kahneman D., Slovic P., Tversky A. (Eds.). Judgment under uncertainty: Heuristics and biases. *Science* 185, 414–421.

Kahneman D. and Tversky A. (1979) Prospect theory: An analysis of decision under risk. *Econometrica*, 47(2), 263–292.

Kahneman D. and Tversky A. (1984). Choices, values and frames. *American Psychologist*, 34(4), 341–350.

Kahneman D. (2011). *Thinking, Fast and Slow*. Farrar, Straus and Giroux.

Kahneman D., Sibony O. and Sunstein C.R. (2021). *Noise: A Flaw in Human Judgement*. Little, Brown Spark.

Kant I. (1797). *Die Metaphysik der Sitten*. F. Nicolovius.

Kaplan A. (1964). *The Conduct of Inquiry: Methodology for Behavioral Science*. Chandler Publishing Company.

Keller E.F. (2000). *The Century of the Gene*. Harvard University Press.

Kelman H.C. (1958). Compliance, identification, and internalization. Three processes of attitude change. *Journal of Conflict Resolution*, 2(1), 51–60.

Keysers C. and Gazzola V. (2010). Social neuroscience: Mirror neurons recorded in humans. *Current Biology*, 20(8), 353–354.

Kjellén U. (2009). The safety measurement problem revisited. *Safety Science*, 47(4), 486–489.

Klein G. (1998). *Sources of Power: How People Make Decisions*. MIT Press.

Köhler W. (1947). *Gestalt Psychology: An Introduction to New Concepts in Modern Psychology*. Liveright.

Komaki J., Barwick K.D. and Scott L.R. (1978). A behavioural approach to occupational safety: Pinpointing and reinforcing in a food manufacturing plant. *Journal of Applied Psychology*, 63(4), 434–445.

Krause T.R. and Bell K.J. (2015). *7 Insights into Safety Leadership*. The Safety Leadership Institute.

Krause T.R., Hidley J.H. and Hodson S.J. (1996). *The Behavior-Based Safety Process: Managing Improvement for an Injury-Free Culture* (2nd edition). Van Nostrand Reinhold.

Kruger J. and Dunning D. (1999). Unskilled and unaware of it: How difficulties in recognizing one's own incompetence lead to inflated self-assessments. *Journal of Personality and Social Psychology*, 77(6), 1121–1134.

Landsberger H.A. (1958). *Hawthorne Revisited: Management and the Worker, Its Critics, and Developments in Human Relations in Industry* (vol. IX). Cornell Studies in Industrial and Labor Relations.

Laporte T.R. and Consolini P.M. (1991). Working in practice but not in theory: Theoretical challenges of "high-reliability organizations". *Journal of Public Administration Research and Theory*, 1(1), 19–47.

Lazarus R.S. (1991). Cognition and motivation in emotion. *American Psychologist*, 46(4), 352–367.

Lederman L. (1992). Debriefing: Toward a systematic assessment of theory and practice. *Simulation and Gaming*, 23(2), 145–160.

LeDoux J. (1998) *The Emotional Brain: The Misterious Underpinnings of Emotional Life*. Weidenfeld & Nicolson.

Leplat J. and Cuny X. (1977). *Introduction à la psychologie du travail*. PUF.

Leplat J. and Hoc J.-M. (1983). Tâche et activité dans l'analyse psychologique des situations. *Cahiers de Psychologie Cognitive*, 3(1), 49–63.

Les Echos. Pourquoi Mark Zuckerberg porte toujours le même t-shirt gris. Accessed from https://www.lesechos.fr/2014/11/pourquoi-mark-zuckerberg-porte-toujours-le-meme-tee-shirt-gris-313358 on April 10, 2025. Published on November 7, 2014.

Lewin K. (1939). Experiments in social space. *Harvard Educational Review*, 9, 21–32.

Lewin K. (1947). Frontiers in group dynamics: Concept, method and reality in social science. *Human Relations*, 1, 5–41.

Libet B. (2004). *Mind Time: The Temporal Factor in Consciousness. Perspectives in Cognitive Neuroscience*. Harvard University Press.

Likert R. (1967). *The Human Organization*. McGraw Hill.

Lovallo D. and Kahneman D. (2003). Delusions of success: How optimism undermines executives' decisions. *Harvard Business Review*, 81(7), 56–63.

Mack A. and Rock I. (1998). *Inattentional Blindness*. MIT Press.

Manser R. and Howard S. (2000). Key elements to debriefing for simulation training. *European Journal of Anaesthesiology*, 17(8), 526–517.

Manuele F. (2003). Heinrich Revisited: Truisms or Myths. In: *On the Practice of Safety*. John Wiley & Sons.

March J.G. (1958). A behavioral theory of decision making. *Personnel Administration*, 21(3), 8–10.

March J.G. and Olsen J.P. (1972). A garbage can model of organizational choice. *Administrative Science Quarterly*, 17(1), 1–25.

Marcus J.H. and Rosekind M.R. (2017). Fatigue in transportation: NTSB investigations and safety recommendations. *Injury Prevention*, 23(4), 232–238.

Markowsky G. (2024). Information theory. Encyclopedia Britannica. Accessed from https://www.britannica.com/science/information-theory on April 10, 2025.

Marquet L.D. (2015). *Turn the Ship Around! A True Story of Turning Followers into Leaders*. Penguin.

Maslow A. (1943). A theory of human motivation. *Psychological Review*, 50, 370–396.

Maust D.T., Lin L.A. and Blow F.C. (2019). Benzodiazepine use and misuse among adults in the United States. *Psychiatric Services*, 70(2), 97–106.

Mayo Clinic. Healthy Lifestyle/Adult health website. Accessed from https://www.mayoclinic.org/healthy-lifestyle/adult-health/in-depth/napping/art-20048319 on April 10, 2025. Published on November 6, 2024.

McCrae R. and Costa P. (1999). A Five-Factor Theory of Personality. In Pervin L.A. and John O.P. (Eds.), *Handbook of personality: Theory and research* (2nd ed.), Guilford.

McElroy M.W. (2000). Integrating complexity theory, knowledge management and organizational learning. *Journal of Knowledge Management*, 4(3), 195–203.

McGregor D. (1960). *The Human Side of Enterprise*. McGraw-Hill.

McGurk H. (1988). Developmental Psychology and the Vision of Speech (McGurk's Inaugural lecture in 1988). Auditory-Visual Speech Processing (AVSP-1988), pp. 3–20.

McLeod S.A. (2015). Skinner – Operant Conditioning. Accessed from https://www.simplypsychology.org/operant-conditioning on April 10, 2025.

Meadows D.H. (2008). *Thinking in Systems*. Sustainability Institute.

Melamed S., Ben-Avi I, Luz J. and Green M.S. (1995). Objective and subjective work monotony: Effects on job satisfaction, psychological distress, and absenteeism in blue-collar workers. *Journal of Applied Psychology*, 80(1), 29–42.

Meyer E. (2014). *The Culture Map – Decoding How People Think, Lead, and Get Things Done Across Cultures*. Public Affairs.

Milgram S. (1974). *Obedience to Authority; An Experimental View*. HarperCollins.

Miller G.A. (1956). The magical number seven, plus minus two: Some limits on our capacity for processing information. *Psychological Review*, 63(2), 81–97

Moore A.H. (2019). Deontological ethics. In: Zalta E.N. (Ed.) *The Stanford Encyclopedia of Philosophy*. Stanford University. Accessed from https://plato.stanford.edu/archives/win2019/entries/hedonism/

National Institute for Occupational Safety and Health. A guide to safety in confined spaces. Cincinnati (OH) DHHS (NIOSH) Publication No. 87–113 (1987).

National Institute on Alcohol Abuse and Alcoholism. Alcohol Use in the United States: Age Groups and Demographic Characteristics. Updated 2024. Accessed from https://www.niaaa.nih.gov/alcohols-effects-health/alcohol-topics/alcohol-facts-and-statistics/alcohol-use-united-states-age-groups-and-demographic-characteristics on April 10, 2025.

Neal D.T., Wood W. and Quinn J.M. (2006). Habits: A repeat performance. *Current Directions in Psychological Science*, 15(4), 198–202.

Nehliq A. (2012). Les effets de la caféine sur le cerveau. *Neurologies*, 15, 153.

Nonaka I. and Takeuchi H. (1995). *The Knowledge Creating Company: How Japanese Companies Create the Dynamics of Innovation*. Oxford University Press.

Occupational Safety and Health Administration. Permit-required confined spaces (PCRS) standard, 29 CFR 1910.146. US Department of Labor; Washington, DC (1993).

Office for National Statistics. Drug misuse in England and Wales. Accessed from https://www.ons.gov.uk/peoplepopulationandcommunity/crimeandjustice/articles/drugmisuseinengland andwales/yearendingjune2022#overall-trends-in-drug-misuse on April 10, 2025. Published on December 15, 2022.

Ombredane A. and Faverge J.-M. (1955). *L'analyse du travail: facteur d'économie humaine et de productivité*. Presses Universitaires de France.

Otero-Millan J., Macknik S.L., Robbins A., McCamy M. and Martinez-Conde S. Stronger misdirection in curved than in straight motion. *Frontiers in Human Neuroscience*, 5, 133 (2011). Accessed from https://www.frontiersin.org/articles/10.3389/fnhum.2011.00133/full

Pashler H. (1994) Dual-task interference in simple tasks: Data and theory. *Psychological Bulletin*, 116(2), 220–244.

Pavlov I. Le réflexe conditionnel. In: Kochtoïanz, K (Ed.) *Oeuvres choisies*, Editions en langues étrangères (1954), translation of an article from I. Pavlov in Grande Encyclopédie Médicale (1934).

Pearson M. and Smith D. (1986). Debriefing in experienced-based learning. *Simulation/Games for Learning*, 16(4), 155–172.

Peltzman S. (1975). The effects of automobile safety regulation. *Journal of Political Economy*, 83(4), 677–726.

Perrow C. (1984). Normal Accidents: Living with High-Risk Technologies. (2nd ed). Princeton University Press (1999). 1st ed. Basic Books.

Perrow C. (1999). Organizing to reduce the vulnerabilities of complexity. *Journal of Contingencies and Crisis Management*, 7(3), 150–155.

Polanyi M. (1983). *The Tacit Dimension*. Peter Smith.

Prayogo A., Teophilea H.S., Nugraha P., Sunindijo R.Y., Maharani C.F. and Yang K. (2024). Noise disturbance increases negative emotions and unsafe behaviour among construction workers. *International Journal of Construction Management*, 19, 1–8.

Radiguet C. Les secrets des quatre phases du sommeil. National Geographic. Accessed from https://www.nationalgeographic.fr/sciences/les-secrets-des-quatre-phases-du-sommeil on April 10, 2025. Published in August 2018.

Rae A.J., Provan D.J., Weber D.E. and Dekker S.W.A. (2018). Safety clutter: The accumulation and persistence of 'safety' work that does not contribute to operational safety. *Policy and Practice in Health and Safety*, 16(2), 194–211.

Rasmussen J. (1997). Risk management in a dynamic society: A modelling problem. *Safety Science*, 27(2), 183–213.

Reason J.T. (1990). *Human Error*. Cambridge University Press.

Reason J.T. (1993). The identification of latent organizational failures in complex systems. In J.A. Wise, V.D. Hopkin and P. Stager (Eds.), Verification and validation of complex systems: Human factors issues. Springer-Verlag.

Reason J.T. (1997). *Managing the Risks of Organizational Accidents*. Routledge.

Reason J.T. (2000). Human error: Models and management. *British Medical Journal*, 320(7237), 768–770. Accessed from https://www.ncbi.nlm.nih.gov/pmc/articles/PMC1070929/

Ricoeur P. (1990). *Soi-même comme un autre* cited in Desjours, C. Le facteur humain, *Le facteur humain*. Presses Universitaires de France/Humensis (1995, 8ème édition 2022).

Riggio R. (2022). Do You Have the Personality for a Leadership Position? Which personality traits are associated with leadership?. *Psychology Today*. Accessed from https://www.psychologytoday.com/us/blog/cutting-edge-leadership/202202/do-you-have-the-personality-for-a-leadership-position

Roberts K. (1989). New challenges in organizational research: High reliability organizations. *Industrial Crisis Quarterly*, 3, 111–125.

Roccas S., Sagiv L., Schwartz S.H. and Knafo A. (2002). The big five personality factors and personal values. *Personality and Social Psychology Bulletin*, 28(6), 789–801.

Rohan M.J. (2000). A rose by any name? The values construct. *Personality and Social Psychology Review*, 4, 255–277.

Rokeach M. (1973). *The Nature of Human Values*. Free Press.

Sanbonmatsu D.M., Strayer D.L., Medeiros-Ward N. and Watson J.M. (2013). Who multi-tasks and why? Multi-tasking ability, perceived multi-tasking ability, impulsivity, and sensation seeking. *Public Library of Science*, 8(1), e54402.

Scharff C. Votre physique influence-t-il votre salaire? L'Echo. Article paru le 28 avril 2010. Accessed from https://www.lecho.be/monargent/travail/votre-physique-influence-t-il-votre-salaire/8909273.html

Schein E.H. and Bennis W.G. (1965). *Personal and Organizational Change through Group Methods: The Laboratory Approach*. John Wiley & Son.

Schein E.H. (2017). *Organizational Culture and Leadership* (5th edition). Wiley.

Schein E.H. and Schein P.A. (2018). *Humble Leadership: The Power of Relationships, Openness, and Trust*. Berret-Koehler.

Scherer K.R., Wallbott H.G. and Summerfield A.B. (Eds.). (1986). *Experiencing Emotion: A Cross-cultural Study*. Cambridge University Press.

Schwartz S.H. (1992). Universals in the content and structure of values: Theoretical advances and empirical tests in 20 countries. In: Zanna M.P. (Ed.) *Advances in Experimental Social Psychology* (Vol. 25, pp. 1–65). Academic Press.

Schwartz S.H. (2012). An overview of the Schwartz theory of basic values. *Online Readings in Psychology and Culture*, 2(1), 11.

Schwarzberg M.N. (1993). Carbon dioxide level as a migraine threshold factor: Hypothesis and possible solutions. *Medical Hypotheses*, 41(1), 35–36.

Schwiedrzik C.M., Sudmann S.S., Thesen T., Wang X., Groppe D.M., Megevand P., Doyle W., Mehta A.D., Devinsky O. and Melloni L. (2018). Medial prefrontal cortex supports perceptual memory. *Current Biology*, 28(18), R1094–R1095.

Seligman C., Olsen J.M. and Zanna M.P. (Eds.). (1996). *The Psychology of Values: The Ontario symposium* (Vol. 8). Erlbaum.

Senge P.M. *The Fifth Discipline: The Art and Practice of the Learning Organization*. Revised edition. Random House Business Books (1st ed. 1990, revised ed. 2006).

Şerif M. (1936). *The Psychology of Social Norms*. Harper and Row.

Shein S.L., Whitticar S., Mascho K.K., Pace E., Speicher R. and Deakins K. (2021). The effects of wearing facemasks on oxygenation and ventilation at rest and during physical activity. *PLoS One*, 16(2), e0247414.

Shepard R.N. (1990). *Mind Sights: Original Visual Illusions, Ambiguities, and Other Anomalies, with a Commentary on the Play of Mind in Perception and Art*. W.H. Freeman and Company.

Simon H.A. (1947). *Administrative Behaviour: A Study of Decision-Making Processes in Administrative Organization* (1st ed.). Macmillan.

Simon H.A. (1956). Rational Choice and the Structure of the Environment. *Psychological Review*, 63(2), 129–138.

Simon H.A. (1972). Theories of bounded rationality. In: Dans Radner R. and McGuire C.B. (Eds.) *Decision and Organization*. North-Holland Pub Co.

Simon H.A. (1996). *Models of My Life*. MIT Press.

Simonetto I. (2020). *Neurosciences et sécurité*. Mardaga.

Skinner B.F. (1938). *The Behavior of Organisms*. Appleton-Century-Crofts.

Slovic P. (1973). Behavioral Problems of Adhering to a Decision Policy. Paper presented at the Institute for Quantitative Research in Finance. Accessed from https://scholarsbank.uoregon.edu/server/api/core/bitstreams/148958f5-1fc5-419d-a56d-8c4583aa4fb5/content on April 10, 2025.

Smith A. (1776). *An Inquiry into the Nature and Causes of the Wealth of Nations*. W. Strahan and T. Cadell.

Snook S.A. (1996). *Practical Drift: The Friendly Fire Shootdown Over Northern Iraq*. Harvard University ProQuest Dissertations Publishing.

Substance Abuse and Mental Health Services Administration. Key substance use and mental health indicators in the United States: Results from the 2022 National Survey on Drug Use and Health. HHS Publication No. PEP23-07-01-006, NSDUH Series H-58. Center for Behavioral Health Statistics and Quality, Substance Abuse and Mental Health Services Administration (2023).

Taleb N.N. (2010). *The Black Swan* (2nd edition). Random House.

Taylor F.W. (1911). *The Principles of Scientific Management*. Harper & Brothers.

Teboul J. and Damier P. (2022). *Le mirage du leadership à l'épreuve des neurosciences*. Odile Jacob.

Thaler R.H. (2016). *Misbehaving: The Making of Bahavioral Economics*. W.W. Norton.

Thaler R.H. and Sunstein C.R. (2008). *Nudge: Improving Decisions about Health, Wealth, and Happiness*. Yale University Press.

Tharaldsen J.E., Olsen E. and Rundmo T. (2008). A longitudinal study of safety climate on the Norwegian continental shelf. *Safety Science*, 46(3), 427–439.

Thiffault P. and Bergeron J. (2003). Monotony of road environment and driver fatigue: A simulator study. *Accident Analysis and Prevention*, 35, 381–391.

Thompson P. (1980). Margaret Thatcher: A new illusion. *Perception*, 9(4), 483–484.

Tissington P., Flin R. (2005). Assessing Risk in Dynamic Situations: Lessons from Fire Service Operations. *Risk Management* 7(4).

Treisman A., Vieria A. and Hayes A. (1992). Automaticity and pre attentive processing. *The American Journal of Psychology*, 105(2), 341–362.

Trobe J.D. (2006). *The Physician's Guide to Eye Care*. American Academy of Ophthalmology, pp. 145–149.

Turner M.E. and Pratkanis A.R. (1998). Twenty-five years of groupthink theory and research: lessons from the evaluation of a theory. *Organizational Behavior and Human Decision Processes*, 73(2–3), 105–115.

U.K. Department for Business & Trade. Business population estimates for the UK and regions 2023: statistical release. Publishd 5 October 2023. Accessed from https://www.gov.uk/government/statistics/business-population-estimates-2023/business-population-estimates-for-the-uk-and-regions-2023-statistical-release

U.S. Centers for Disease Control and Prevention. Sleep and Sleep Disorders; FastStats: Sleep in Adults (2024). Accessed from https://www.cdc.gov/sleep/data-research/facts-stats/adults-sleep-facts-and-stats.html on April 10, 2025. Published on May 15, 2024.

U.S. Small Business Administration, Office of Advocacy. What's New with Small Business, Frequently Asked Questions. Mars 7, 2023. Accessed from https://advocacy.sba.gov/2023/03/07/frequently-asked-questions-about-small-business-2023/#:~:text=There%20are%2033%2C185%2C550%20small%20businesses,net%20jobs%20created%20since%201995

Underwood G. (2005). *Traffic and Transport Psychology: Theory and Application. Proceedings of the ICTTP 2004*. Elsevier. pp. 455–456.

UNESCO. Déclaration de Mexico sur les politiques culturelles. Conférence mondiale sur les politiques culturelles, Mexico City (1982).

United Nations Office on Drugs and Crime. World Drug Report 2023. Accessed from https://www.unodc.org/res/WDR-2023/Special_Points_WDR2023_web_DP.pdf on April 10, 2025.

Vaughan D. (2016). *The Challenger Launch Decision: Risky Technology, Culture, and Deviance at NASA* (enlarged edition). University of Chicago Press.

Veblen T.B. *Why Is Economics Not an Evolutionary Science?* (1898). Reprinted in The Place of Science in Modern Civilization (1919).

Von Helmholtz H. *Handbuch der physiologischen Optik.* Vol 3. (1910). Translation by The Optical Society of America's Southall (1924).

Wacongne C., Changeux J.-P. and Dehaene S. (2012). A neuronal model of predictive coding accounting for the mismatch negativity. *Journal of Neuroscience*, 32(11), 3665–3678.

Wandell B., Dumoulin S. and Brewer A. (2007). Visual field maps in human cortex. *Neuron*, 56(2), 366–383.

Warm J.S., Parasuraman R. and Matthews G. (2008). Vigilance requires hard mental work and is stressful. *Human Factors*, 50(3), 433–441.

Wason P.C. (1960). On the failure to eliminate hypotheses in a conceptual task. *Quarterly Journal of Experimental Psychology*, 12(3), 129–140.

Watson J.B. (1913). Psychology as the behaviorist views it. *Psychological Review*, 20, 158–177.

Weber M. *La domination (1ère éd. 1914).* French edition by Kalinowski, I. La Découverte (2015).

Weick K.E. and Sutcliffe K.M. (2001). *Managing the Unexpected: Assuring High Performance in an Age of Complexity.* Jossey-Bass.

Westrum, R.A. (1984). Complex organizations: Growth, struggle, and change. Prentice-Hall.

Westrum R.A. (2004). A typology of organizational cultures. *Quality Safety Health Care*, 13(Suppl II): ii22–ii27.

Wicker A.W. (1969). Attitudes versus actions: The relationship of verbal and overt behavioral responses to attitude objects. *Journal of Social Issues*, 25, 41–78.

Wilde G.J.S. (1998). Risk homeostasis theory: An overview. *Injury Prevention*, 4, 89–91.

Wilson L. Safestart. Accessed from https://fr.safestart.com/wp-content/uploads/2018/11/SafeStart_2018_Les-Facteurs-Humains_Article_FRE.pdf on April 2025. Published in 2018.

Wilson T.D. and Schooler J.W. (1991). Thinking too much: Introspection can reduce the quality of preferences and decisions. *Journal of Personality and Social Psychology*, 60, 181–192.

Woollet K. and Maguire E.A. (2011). Acquiring "the knowledge" of London's layout drives structural brain changes. *Current Biology*, 21(24), 2109–2114.

World Health Organization. Blindness and Vision Impairment (2023). Accessed from https://www.who.int/news-room/fact-sheets/detail/blindness-and-visual-impairment on April 10, 2025.

World Health Organization. Sensory Functions, Disability and Rehabilitation Team (2019). World report on vision. Accessed from https://www.who.int/publications/i/item/world-report-on-vision on April 10, 2025. Published on October 8, 2019.

Young M.S., Brookhuis K.A., Wickens C.D. and Hancock P.A. (2015). State of science: Mental workload in ergonomics. *Ergonomics*, 58(1), 1–17.

Zac P.J. (2017). *Trust Factor: The Science of Creating High-Performance Companies.* AMACOM.

Index

Note: *Italic* page numbers refer to figures.